FULL ECOLOGY

Repairing Our Relationship with the Natural World

Mary M. Clare and Gary Ferguson

Heyday, Berkeley, California

Library of Congress Cataloging-in-Publication Data
Names: Clare, Mary M., 1957- author. | Ferguson, Gary, 1956- author.
Title: Full ecology : repairing our relationship with the natural world / Mary M. Clare and Gary Ferguson.
Description: Berkeley, California : Heyday, [2021]
Identifiers: LCCN 2020041931 (print) | LCCN 2020041932 (ebook) | ISBN 9781597145183 (hardcover) | ISBN 9781597145190 (epub)
Subjects: LCSH: Human ecology--Philosophy.
Classification: LCC GF49 .C59 2021 (print) | LCC GF49 (ebook) | DDC 304.201--dc23
LC record available at https://lccn.loc.gov/2020041931
LC ebook record available at https://lccn.loc.gov/2020041932

Cover Photo: Julian Piehler
Cover and Interior Design/Typesetting: Ashley Ingram

Published by Heyday
P.O. Box 9145, Berkeley, California 94709
(510) 549-3564
heydaybooks.com

Printed in East Peoria, Illinois, by Versa Press, Inc.

10 9 8 7 6 5 4 3 2 1

For our grandmothers, Mary and Etta.
And for that inspiring grandmother grizzly of
the northern Rockies, #399.

It's not "Wherever I stand belongs to me."
It's "Wherever I stand, I belong."
–Bruce McQuakay
Tlingit/Haida, Klamath/Cree

———————

Then you turn around to find peace
in the last place you've thought to look,
the natural world deep inside you.

TABLE OF CONTENTS

INTRODUCTION

This is a photo of Mary and her just-younger sister, Alicia. Busy preschoolers captured on Kodachrome in the midst of their first encounter with tadpoles. Tipsy with the mystery of those wiggly beings, the girls reach, listening to the promise their mother is offering. "We can come back next week, and maybe the week after, to watch them turn into frogs." It's an astonishing idea: the unfolding of frogs. Still, in that moment, ideas are too weighty. Mary and Alicia are tuned into a perfect mix of things, just being. Little girls, nascent frogs, the chill water of an alpine pond in early summer—all nestled under the bright drape of sky and walls of granite keeping watch over their home in the Yosemite Valley.

Look at all these birds!

This is a photo of Gary, his brother Jim, and their dad, on an untrammeled stretch of Lake Michigan shoreline. Gary is three years old. In one hand he's holding pieces of bread to feed the seagulls. But in this moment, he can't quite focus on the hungry birds. He's mesmerized. Endless waves roll to the shore with no one pushing them. The gulls circle and then disappear over the dunes, only to be followed by another group. And all that sand, stretching as far as he can see. His parents tell him the beach was made of tiny pieces of broken rock. But who broke the rock? The mysteries are stacking up.

You, too, likely have memories like this from your childhood. Tucked away, but still well within reach. Indelible imprints of the wide world beyond your small body. You in the rain, the wind, the snow. Wandering through the forest, walking into the waves. Catching bugs in the yard, climbing a tree in the park, listening to the wind singing between tall buildings.

Hold these memories as you read on. They are calling cards to the wholeness of a human lifetime, strongly rooted in your nature. Guiding lights showing the way back to the landscape of your being. A homecoming, as uncomplicated and as mysterious as tadpoles in an alpine pond or seagulls dancing above the dunes.

Some days, and lately more than any of us would like, the easy joy of childhood can seem out of reach, especially smack in the middle of advancing climate change. The forecasts of intense weather and food insecurity, polar melting and skyrocketing releases of methane are unrelenting. It's clear that our species is largely responsible, and that none of this trouble is going away on its own. With no time to waste we know we need to act, right now, to correct the damage. Yet, smart as we often are, we're stuck. We don't know where to begin, but it's clear that we must.

So, we calm just long enough to see that we can only ever begin from where we are. And from this very place, the first thing to do is to slow down, to listen.

You can start in the next few hours by intentionally shifting even twenty minutes of your attention from the blue light of your screens and the siren song of the twenty-four-hour news cycle to the wide world all around you. No matter where you are, stop—listen, look, smell, touch. Neuroscience has shown that this simple shift in focus, especially when you do it consistently, actually rewires your brain, reorienting you to the living reality in which you're immersed. This simple but profound shift helps you act in ways that support not only your own health, but also the health of the planet you call home.

Think right now: If today were your last day and you knew it, what would matter most? Really. Right now.

This is the question climate change places right at your feet. It's not going away and it will radically shift your priorities. It will change what you need. What you prize. The way you are with other people.

The book in your hands can't wrap in a neat bow some set of finite steps guaranteed to renew climate health. If it were that simple, we'd be a lot further along by now. Instead, this book calls you to pay careful attention to the intelligence of the natural world—both inside and outside of you. The truth is you already have a brilliant inner compass with a needle that reliably aligns to the exquisitely creative, most trustworthy aspects of nature. In this way you find the path forward, stepping together with more companions than you can imagine, into the heartfelt work of repair.

Full Ecology started putting down roots when the two of us met, late in the spring of 2013. Up to then we'd lived nine hundred miles apart, birds of different professional feathers, traveling on separate flyways. Mary was in Portland, Oregon, where she'd spent thirty years as a professor in the social sciences, writing and preparing graduate students to become public leaders in education, mental health, and community government. Meanwhile, from his home in southwest Montana, Gary had been making wide-ranging forays into the planet's wild places, telling of them in a stream of books and essays.

Looking back, our meeting seems against all odds. Hunkered down in the bone-chilling middle of a dark, wet Portland spring, Mary made it as far as spring break before deciding to "leave the field." After decades of pledging to visit her old college friend Joe—to finally meet his wife and their two girls—she got in the car, drove through rain and snow and over four mountain passes all the way into the middle of Montana. Late that night, closing in on her destination, the clouds parted to reveal a full moon rising over the Ruby Mountains.

As it happened, Mary's friend Joe, a truly gifted guitarist, had discovered Gary's knack for playing the blues harp; together with a couple of friends they'd formed a grown-up garage band, christening it with the somewhat dubious name of Arroyo Speedwagon. Anticipating Mary's visit, Joe had asked Gary to come for dinner, saying only that a "buddy from college" was in town. And so, by a big splash of dumb luck and a good measure of divine intervention the two of us found each other. The pizza was mediocre. The conversation was electrifying.

So began talks that would roll out for years, and mostly while we were side by side, walking. Wandering mountain trails and rural backroads of Montana, or threading miles of Portland's forested city streets. In some ways it was like riffing in jazz; we kept finding rich connections, one after another, between Gary's world of conservation science and Mary's world of psychological and cultural studies.

Early on, Gary would tell Mary of the hours he'd spent watching the Yellowstone wolf packs—their easy, yet sophisticated cooperation as they traveled great distances to find prey, then returned home together to care for pups and elders. These tales would spark Mary's stories of cooperation among humans. Like how she'd watched the leaders of seventy-two distinct Native American Tribes and Canada First Nations bring to a close a decades-long negotiation, reaching full consensus on standards for protecting the water of the Yukon River Basin. Whether in nature or among humans, cooperation carries life forward.

As the months rolled by, more moments of synchrony unfolded. Ambling beneath the London plane trees in Portland's Laurelhurst Park, Mary described her work with migrant families; in particular, how building alliances between parents and school administrators led to increased success in children's learning. Gary riffed in with a description of how adult elephants join together to ready a young female for entering the role of matriarch for the herd. Another time, walking the surf line of a quiet beach in Baja, Mary spied a pod of whales. She'd been thinking about challenges of addiction and trauma; how, for humans, social connection is key to healing. As we watched, a mother and baby orca breeched in tandem, leading Gary to recall recent findings showing how adult and adolescent orcas draw together to support the orphaned and injured. From whales to elephants to chimpanzees to wolves to humans, community keeps us healthy.

So it went, and as the miles stacked up, we came to see with growing clarity how factors that lead to health in nature are mirrored in humans. We started understanding that natural ecologies and social ecologies are, in fact, vigorously linked. We've come to call this set of connections *Full Ecology*.

Full Ecology shines a light on the fact that to be human is to be in and of the natural world—a reality most of us rarely consider in passing, let alone by coming to a full stop. As a result, we keep missing the wisdom, the direction to be found there. Up against the crisis that is climate change—its unrelenting cascade of catastrophes impossible to undo—we need this wisdom.

It's not too late. Or, said another way, it's only too late if we keep going forward as we have, unconscious of the rules and beliefs we live from—and more specifically, how those ways of knowing and being contribute to the problem.

This is what makes Full Ecology *full*. Instead of ignoring the thoughts and beliefs that drive us, we aim for thorough and honest consideration of our social ecologies—the ecologies between us and within us. Like biological ecologies, which thrive from active relationship among all living things, social ecologies are the infinite points of interdependence between you and the life that sustains you. The way you interact with the rest of the world, human and otherwise, depends profoundly on who you *think* you are. These beliefs filter what you let in and what you let out. They determine what moves or comforts you and what makes you afraid. They affect decisions about where you live, what you buy, who you love and how you work. More publicly, social ecologies mediate every matter of justice and injustice, every policy on healthcare, and childcare, and the environment. Social ecologies are where human actions originate. And they can either invigorate or degrade our chances for living worthy lives on a wounded planet.

This is a book about repairing the bridge between the ecology "out there"—the healthy functioning of a forest or ocean or grassland—and the health and creative capacity of your thoughts and perceptions. It suggests an important question, one we'll be coming back to again and again in the coming pages. *In any given moment, are you at all separate from nature?* This question is a star to steer by. A question that, when fully engaged, will reveal the intricacy and reliability with which the wisdom of the natural world lives in *you*.

You can actually *choose* to more fully align yourself with natural processes. And the more you choose it, the smoother the choice becomes. What we're called to do is embrace the deep biological reality of our connections with nature, while at the same time understanding that our best, most life-affirming actions flow from the relationships that sustain those very connections.

Chinese ancients observed that "only fish cannot know water." Water is everywhere, the whole medium of their lives. As far as the fish are concerned, water isn't a thing at all. But what fish don't know doesn't hurt them; the water sustains them whether they can see it or not. Here in the human world, though, we've taken to swimming in an invisible medium of a different sort—one made up of unseen ideas and assumptions that have, over time, become dangerous.

Continuing to swim around with such limited vision could kill us. That's why you're here. You already have a healthy suspicion that there's something beyond these waters. You're right. Consider for a minute how most of us don't spend much time thinking about breathing. Yet not one single thought, feeling, or action happens without breath. Breath is how you gather oxygen to sustain the roughly thirty-five trillion cells that make it possible for you to think and walk and laugh. Take that a little further: The oxygen you need to survive is only available because plants, including tiny ocean plants called phytoplankton, are busy making a living by moving carbon and sunlight through their bodies to generate it. Those plants breathe, in turn, by taking in the carbon dioxide that you and countless others exhale. And breath is just the beginning. The phosphorus your body needs to repair tissues and strengthen your teeth travels from the soil and into the cells of your body through the plants you eat. And there's more. Simply being near a forest can lower your cortisol (stress hormone) levels, while the infrared sunlight that bathes you each day sets the melatonin levels you need to go to sleep at night.

There's no getting around the fact that you're in very deep with the natural world. Every single minute of every day. That means that any thought system that narrows this relational reality is by default incomplete. The growing urgency of climate breakdown has left many of us hungry for bigger ideas, larger visions. And few visions are bigger, or have more heart and consequence, than your beautiful, sprawling, mysterious relationship with our home planet.

Again, it's all too easy to overlook the limits of the pond in which you swim. When that's the extent of your vision, it's impossible to see the meadow beyond the banks of the pond, or the forest and mountains beyond that. Learning to see those things takes developing whole new ways of knowing – new and larger understandings of what's real. And that requires crawling out of the water of the world as you know it, onto the mud and grass. It requires the courage to enter the unknown, to look and to learn. It requires the courage to change. Full Ecology is about using our human nature, our powers of attention and self-reflection, to expand what we know. In turn, that wider perspective suggests new ways to act. Think about the perseverance of a hatchling box turtle as it ventures first to break through its eggshell, then to make way through the layers of earth beneath which its egg is buried. Drawing on its nature, inherited from fifteen million years' worth of ancestors, the hatchling develops a carbuncle—a temporary egg tooth that makes it possible to break open the shell. Once free, the young turtle must begin digging—a prolonged and incremental effort that, after three to seven days, allows it to push all the way through the depth of sand, and break into open air.

You have your own versions of carbuncles for opening yourself to new ways of perceiving your relation to the world around you. As with the fledgling box turtles, you'll need persistence. One step and then another. Moving ever forward until your inquiries, your tireless small actions lift you up and onto the ground, finally to look into the world beyond.

––––––––––––––––––––––

Evolutionary biology suggests that organisms are always moving toward a self-transcendence that leads to something not just more complex, but also more adaptable and resilient. It's been this way since the very beginning. Atoms organized into molecules that in turn ended up joining together across a thousand steps to form cells. From those first cells would eventually come all the beings we see on Earth, the vast majority of whom have been in continual tweaking mode to better meet the challenges of changing circumstances. Success for these organisms comes from the fact that they contain and rely on every single thing that came before, including the ability to thrive by being in full connection with everything around them.

Life, then, results from both individual agency and vibrant interconnection. For fish to wriggle onto the land and become terrestrial required constant collaboration between the collective forces of individual acts and complete interdependence. And here's the exciting thing: we humans, rich with the gift of reflective thinking, are actually able, if we wish, to *direct* our participation in this magnificent bit of choreography, consciously applying our individual agency to fortifying our lifelines within the intricate web of relationship.

If the brilliance of nature is all around us—and further, if there's similar brilliance in each of us—why doesn't it show up more fully, more clearly, as a guiding force? Where has that guidance been as we've soldiered forth on countless offensives into productivity, politics, and procreation? Why, with our well-being stuck in the jaws of climate breakdown, is our own natural integrity and ingenuity so hard to see, let alone to unleash? The short answer is that a large and dominant portion of our species has become mired in a very particular way of seeing the world—something we call *separate thinking*. Over thousands of years dominant cultures have become increasingly fascinated with what appears to be outside of us. This has led to the belief that humans are separate from each other and from the natural world. Without really knowing what we were doing, we created a separation myth, and it's severely hobbling us.

Across your lifetime you've likely been conditioned through virtually every social group you've been a part of, every institution, to perceive yourself and the land you live on as a tumble of distinct objects. Separate thinking so dominates communication and exchange across the planet that, for most of us, it's completely reflexive. As invisible as the air we breathe.

Not that such reductionist thinking is useless. In fact, it's a really good tool for revealing a whole host of partial truths. Like the way leaves make energy from sunlight. Water always freezes when its temperature dips below a certain point. Gravity makes spilled rice fall to the floor; then when it hits, the force of the solid surface scatters it all over the place. Knowing the how and why of such things has allowed us to predict them, and then figure out how to use that knowledge to our benefit.

Contained facts like these are fine. We rely on them daily. But by themselves, they never tell the whole story. Thinking in terms of separation, considering things only in isolation, is always incomplete. It's like knowing that an acorn needs to be planted in the dirt but not knowing that it needs water and sun and carbon dioxide to become an oak tree. By dedicating our lives to limited truths without attention to the complex weave of relationship, separate thinking becomes the water we swim in. And as a result, social ecologies become dominated by the separation myth—by individual expressions of power *over*, rather than power *with*. Power over the land and its resources. And power over other people.

When our social ecologies are overwhelmed by separate thinking, it leaves us knowing ourselves first and foremost as lone objects in a wild and wooly universe of objects. That's a scary place to try to live. Isolated in separate thinking, we're afraid, prone to being insular, too easily knowing ourselves by who we're *not*. Not the person who speaks another language, has different color skin, a different level of education, a different way of loving, another way of praying. And beyond that, well, it's pretty obvious that I'm not related to a tree, or a koala bear, or an elephant.

At which point we need a reality check. No place on Earth has there *ever* been such a thing as a rugged individual. Nothing in nature, including humans, ever exists in singularity. In these real and pressing days of climate breakdown we're being called to muster the courage and humility to stop—first to see, and then to step beyond exclusive separate thinking and into the actual *realities* of life. It's time to marry our individuality with the relationships that forged it in the first place.

Holding still to really look, to take in new information, to rethink what's real, will change you. Resting at the pond's edge, you can't help but move into a different understanding of the water that used to be your whole world. It's a kind of epiphany, this realization that the pond is just one part of the story.

Fortunately, self-reflection is central to human nature. But using it well in the face of climate change requires breaking the trance of some very deeply socialized routines. To be completely honest, stepping onto land and into the expansive homeland beyond the edges of the pond also requires courage. There's big country out here, loaded with uncertainty and unknowns—but also far more real.

Thinking *with* nature instead of outside of it calls for two deliberate commitments. First, a commitment to calm down—to stop, to step away from reactive impulses to *do something;* or for the more passive among us, to react in the other direction, becoming paralyzed in worry. Second, thinking with nature means seeing how each of us, and thus our culture as a whole, are shackled by our habits of acting out of separation.

Be aware that in this very moment you might be unleashing a few well-practiced thoughts. *There's no time for this. Slowing down, when the planet is unraveling? I don't think so!* We have these thoughts ourselves. Yet we've become convinced that stopping to see clearly where we are, then proceeding slowly enough to fully understand what's gotten us into this mess, is the most efficient way to get to climate restoration and repair.

We've learned this from nature itself: The elephant matriarch stops the everyday routine of the herd to sense the tipping point in a drought season, realizing it's time to start a long journey to water; the leaders of a wolf pack hold still to size up an elk herd before starting the hunt. Nature is focused always on the realities of the present. It tweaks itself into vitality much like a sailor who reaches port by constantly fiddling with the tiller, deeply connected to the reality of each moment.

As Full Ecology has come into focus, we've stopped to recognize the way our beliefs frame what we can see of the world. We've watched those frames shatter with new information, leaving us to build bigger ones, all along fine-tuning what we know and how we see. As tireless researchers (aka nerds), we turn regularly to the findings of human development, neuropsychology, cultural psychology, and to those areas of natural science related to healthy ecosystems. We've investigated intellectual history, cosmology, and spiritual traditions to distill the most important thing you'll find here: ways you can engage right now to live more fully and effectively in the press of climate change.

The shift we're talking about takes more than "getting it" in theory. The vitality and viability of this change depends on what you admit, what you come to know, and then on what you do. In effect, the challenges of climate change are asking us to grow up: to tell the truth, trusting that we have everything we need to advance into a larger way of knowing and being.

Human history is filled with countless historic moments when significant numbers of people rejected, revised, and moved beyond destructive and even obliterating cultural narratives. In the spirit of "the hundredth monkey," researchers have found that tipping points leading to sweeping cultural change often kick into high gear once about 10 percent of the population decides there's a better way. During medieval times, for example, western Europe was on the tail end of a thousand years spent policing people with harsh rules: *I'm right, you're wrong. I live, you die.* By the seventeenth century, however, an era known as the Enlightenment took hold. And while much oppression remained (and still remains), there came in this time a powerful "10 percent shift" in which large numbers of people began embracing the idea of universal human value—a value we still struggle mightily to enact.

Fast forward to our time, to right now, when our collective task is one of growing up. No matter how risky or inconvenient, it's time to do what it takes to step up and show up in the next 10 percent. Interdependence is our life source. The changes that any of us makes individually will influence the way we think and act collectively.

The circumstances of climate change call our hand. It's time to see just how devoted—even addicted—we are to separate thinking. Climate breakdown demands that we transition our attention and actions in ways that assume full and active responsibility from our interdependent place in the world. That we act in ways fully consistent with who we are: *human nature* inseparable from all of nature.

You *are* nature. And even though nature's survival doesn't require the presence of any of us humans, we all belong. And we are thoroughly connected. At this point, reclaiming and living from that reality is what you can do to be one among the 10 percent—part of the solution instead of the problem.

Full Ecology is based in the recognition that revising our relationship with nature isn't just overdue; it's urgent. In service to that shift, we've organized this book around four fundamental human behaviors. We believe that these behaviors, when consciously engaged, can have immediate, enduring impact on climate repair. *Stop, Ask, Act, Inspire.* Each is nourished and refined by self-reflection, and at the same time, each is grounded directly on the wisdom of interdependence.

First, we *Stop.*

Beyond the panic and fury over climate breakdown lie habits of knowing and acting that degrade our species and our planet. Being mostly unconscious, those habits cannot be changed without stopping long enough to see them. Stop doing. Stop doing, stop trying, stop seeking. Stop so you are able honestly to describe the circumstances before you.

The next step is to *Ask* what's really going on. What circumstances are sustaining the problems unmasked when you come to a stop? As you ask, you learn more about the source and structure of a problem; and along the way you get better at formulating questions that will shape your future actions. Paradoxically, slowing to ask leads to better-focused and far more efficient actions.

The third behavior, then, is to *Act*. Having figured out how you're making sense of the world allows clarity to emerge. And through that clarity comes guidance, moving you into actions that are responsive, self-correcting, and in the case of climate change, healing.

Finally, acts of healing and connection can't help but lead to the last step on this four-part path—to *Inspire*. You know what it's like to be around somebody inspiring. They shine a light by virtue of a deep integrity between what they say and what they do. Their lives make you want to pay attention—to learn ways you too can act with integrity. Healing actions can nudge us toward a powerful shared resolve. Acts of repair gain strength as more and more people commit to the web of our connection.

You've likely come to this book knowing too well some of the ill-advised, unsustainable ways of humans. You see the damage, not only to people but to the oceans and the air, to the forests and the rivers, to the animals and the plants. Sometimes it's baffling that these facts don't lead to big and immediate changes. And quick changes *are* possible. The problem is that they're not very likely to take hold and then have staying power. Essential changes in human awareness and behavior take time.

Full Ecology offers a way to be actively in sync with the time, the courage, and the intelligence this kind of change requires. It gives you a set of steps for reclaiming and activating the fullness of your nature. It helps you uncover what you need to upgrade the story you tell about being human. The point is to meet climate change with your deepest, smartest humanity—to act from your innate instinct for care and connection to help correct our course. We have everything we need, as close as breath, as reliable as life itself.

STOP

It's hard to overstate how much we prize quick, decisive action. We perform our public lives as action figures—heroes starring in our own suspense-filled movies, with riveting plans tucked into our pockets. Clearly, there are times when heroism is just the thing. Your child spikes a fever, and you spring into action to get her to the doctor. A young doe dashes out onto the highway and you deftly brake and swerve so that everybody lives. If a wildfire threatened your home, you'd gather your family, grab the few things most urgently needed, clear out, and then do what you could to check on the safety of your neighbors. At such times, fast action is just good sense.

Then comes along an unprecedented, hugely tangled problem like climate change. With every passing season, breakdown takes firmer hold, and at a rate considerably faster than science predicted. Your sense of urgency and impulse to action ignite but you're pretty sure flash responses won't solve much. Early on, you might have looked to "them," the authorities somewhere out there who ought to be on top of things. Increasingly, though, no one is on top of things, and you find yourself casting about for what to do. You walk or bike more often, put on a sweater and lower the thermostat in winter. You buy organic or drive a Prius. But it never feels like enough. At the periphery, anxious questions run circles in your mind. *How can anyone keep going? Or even stay sane? Don't they get it that this planet is suffering? What in the world can I do?*

Situated as our species is, right here and now in the web of life, Full Ecology turns to nature for answers. Here in nature we've found two strong and immediate truths that can guide our efforts. First, nature cultivates and nourishes connection, a quality that has helped it sustain itself as a thriving garden. Second, nature manages to be both supremely methodical and highly improvisational. It reacts to changes with agency and relationship, activating time-tested strengths to craft fresh responses in every moment.

And guess what? Given that you're no less natural than a redwood or a grizzly bear or a hummingbird or an orca whale, you have access to these exact same powers. But without stopping long enough to see, right now, where you and our species are going wrong, you're unlikely to draw on your connection with all life, or to fully activate your skill for responding well. To do this requires engaging the observational abilities nature has given you. Central to this is the fact that you think a lot about what you observe. This is a huge asset. It's also a liability. You'll see why.

When it comes down to it, thinking is the primary way you come to have any idea of what's real. You use your thinking to assemble the stories you consult as you make your way through your days. Stories of what's safe to eat and what isn't, stories of who you vibe with as friendly and who you avoid. Among all these stories are the things you tell yourself about the natural world, and those things determine how you live with and in it.

You always have opportunity to develop and change what you know—to pay attention in ways that refresh and revise your stories with wider understandings, bigger views. Nature invites this kind of growth through kinship. The natural world is with you all the time. You always have the choice to turn to the natural world for guidance when life gets too tangled, when next steps are unclear. Yes, you're dealing with very different circumstances than those of a lion or a redwood tree. And yet, nature's ways of staying strong and creative are part of you, too. From your kinship with nature, you can learn to apply better ways of being human. Reclaiming your human nature is all about recovering and expanding your range of reliable methods and exquisite improvisations.

Because of the way your brain naturally works, however, coming into alignment with your truest nature has a serious prerequisite. You have to stop. Full stop. Stop everything. Stop planning and managing. Stop long enough to see that all the ideas you have are incomplete. Stop to feel the connection, the brilliance unfolding in and around you all the time. Give all your attention to how, all along, you've belonged here, this wide world making possible everything you've ever done. Consider how you've kept that brilliance out of reach.

We're a species in furious motion. Unless we stop, we can't see how we're continuing to behave in the ways that created climate breakdown in the first place. The way out is through relationship, through our natural kinship with all things. But first we have to admit that we're continuing to obscure and damage those connections.

Learning theorists have found that well beyond what gets measured by standard IQ tests, humans possess at least a dozen kinds of intelligence—among them, emotional, social, musical, kinesthetic, and moral. This first touchstone of Full Ecology, to *Stop*, is about clearing the way for growing your *ecological intelligence*. By this we mean the intelligence of stewardship, the intelligence that shows up to guide how you sustain and care for your relationships with all things, animate and inanimate, human and otherwise—*all things*. To stop—to stop everything—lets you taste the fullness of your relationship, your connection. It lets you feel how that connection is not just a concept or belief, but instead the bedrock of what is real.

All of that may sound downright poetic. But it's also hugely practical. Much as a healthy oak tree stands open, in every minute adjusting in relation to sun and soil and microbes and rain, you are in your highest expression of ecological intelligence when your thoughts and actions are in sync with the web of life pulsing in and around you.

Especially in a culture like ours, with feeds from social networks and stock markets and advertisers screaming for our attention, stopping to get the lay of the land is critical—and, when you think about it, radical. In the midst of these everyday maelstroms, *stopping is actually doing something*. Stopping makes space for seeing faulty thinking, for undressing the elaborate filters that shape your perceptions of nearly everything. Stopping releases you from the partial truths you've used to stake down your life. It's a big and necessary step in preparing the ground from which more creative, more satisfying, and more sustainable actions can sprout and grow. Stopping opens a channel to all the energy you'll need to keep going over the long haul. It is vital to your ecological intelligence.

Stop and watch where you're putting your attention. This might sound easy, but tracking the link between your thoughts and the version of reality they come from takes a kind of discipline different from what you might expect.

Take a look right now. You're paying attention to the words on this page. You're probably also having thoughts beyond these words. Some of those thoughts serve to make what you're reading more personal. You might think something like, "Hmmm. I really don't pay much attention to my attention." On the other hand, you might think, "Who has time to stop?" You might also have an unrelated thought like, "I really can't forget the oil change I scheduled for today." You know how it goes.

The takeaway here is that when you don't pay attention to your attention, you forfeit the chance to make conscious choices about where to put it. Attention is one of the most precious abilities you have. It's what's given you your worldview—your collection of beliefs about what's real. And from here forward, what you attend to will either stamp in or modify those beliefs.

Try this short exercise in stopping. All you need is five or ten minutes, a pen, and a piece of paper. Over the years, life has no doubt given you some of what you'd hoped for, while other things have remained elusive, out of reach. For this exercise, simply write down items from that latter group—things that today you really, really want. Go for it. Not even the sky is the limit. You can be as specific or as esoteric as you like. It's your list. For your eyes only.

When you've finished, scan what you've written. Give it some thought, and when you're ready, circle the one thing on the list that you want the very most. There's no wrong answer. Now, with no praise or blame, read that one thing aloud to yourself. For right now, at least, this is what your life serves. It's where you place your most heartfelt attention.

Maybe the thing you circled was a red Maserati. But probably not. Maybe you found that the thing you want the most is a big, generous wish. Something true as true can be that you want with all your heart. Something like "world peace" or "a healthy planet." If so, and if you're like most of us, after a beautiful minute or maybe two spent holding the promise of that desire, something dark can yank you sharply back to what you're used to believing is real. And just like that, the air goes out of your lovely wish.

Let that feeling be your cue to press pause again.

Your wish didn't deflate because it ran up against some jagged reality. Your desire for world peace, or a healthy planet, deflated because you live in a society hooked on *the product*, with little tolerance for vision without an airtight logic trail. Stop to see what a bad habit that is. Deeply remember that the worth of a desire never depends on immediately grasping the mechanics of achieving the final product. This isn't to deny the value of results. But if nothing makes it into your heart's desire except what can be packaged and measured right now, you'll pass up some spectacularly valuable directional signs for living your truest life.

Stopping reveals the essential truth of wishing for, say, world peace. Peace is a desire that keeps your interconnectedness at the forefront of your heart and mind. It causes the illusion of separation to shrink just a little bit, leaving relationship to play out in creative ways. In the same way, your wish for a healthy planet lives in connection. If, when pressed by the pragmatic skeptics in the room, you conclude that your dream is hopeless, you again shore up the status quo of the separation myth—for them and for you. Everyone gets to shake their heads, toss out a few cynical curses, and retreat to their familiar nooks of separation when, in truth, peace and planetary health are vital to sustaining life.

And know this: *All of nature—all of it—seeks sustainability.* Species that disconnect or adopt other destructive ways end up leaving the dance. If, in their eager quest for food, thistle-lipped tweeter monkeys (a species we made up) start gobbling up patches of death cap mushrooms, that is arguably entirely natural. It's also natural for them to not be long for the planet. In fact, as nature ourselves, the choices we've made that have brought us to climate disaster are expressions of nature, too. The centuries-long domination of exclusive separate thinking is indeed natural. As is the resulting chaos in climate breakdown, and in global explosions of violence, poverty, war, and oppression of all kinds.

Stopping makes it possible to see your own participation in ways of thinking and acting that, like the tweeter monkey's impulsive feasting, are driving our demise. That said, this kind of pausing requires real courage. When you stop long enough to really see your separate thinking, it can feel like you're going against what's always worked. As any recovering addict can tell you, it's no small thing to see and admit habits of thought and action that were meant to keep you safe, but really don't. The possibility for breakthrough with a full stop can come right alongside profound feelings of imbalance, like a rug being jerked from beneath your feet. But stick with it. Sooner than you imagine, you'll realize that stopping also places right in your lap the quiet clarity to see the relationships that make life worth living—things that matter far more than stubborn adherence to cherished illusions.

For most of us, it's not far into any thought stream before the idea of *me* arises. Me reading. Me thinking. Me worried about the climate. Once you see how quickly the *me story* kicks in, you'll be right on top of the separation myth. There's not really anything wrong or weird about *me* thoughts. They make all sorts of sense given our distinct experiences of life in a body. But *me* stories are woven out of separate thinking, and when that kind of thinking is the only game in town, well, there's trouble.

Watch how the *me story* behaves. Sometimes you can't really see it until you notice how easily it hides in a *not-me story*—the story of other, of opponent. Once you watch for that, you'll see it everywhere. Fox News versus MSNBC. Republican versus Democrat. Christian versus Muslim. For almost every one of us, separate thinking is the default anchor for our attention. That's a big deal. Because where we place our attention determines whether we continue in climate decline or find ways to effect climate repair.

When Gary was a boy, growing up in the fading industrial belt of the Midwest, he found himself obsessed with wild places—with the uncommon freedom and at-homeness he sensed there. At nine years old he announced to his parents that he was going to live in the Rocky Mountains; four years later, he strolled into the living room with road maps in hand to tell them he'd saved enough money from mowing lawns to ride his stingray bike to Colorado.

On one hand, Gary wanted to experience the thrill of the West's unshackled wild places. But he also longed to escape the exhausted, polluted lands of his boyhood. They were two very different, and to him thoroughly opposing, worlds. He eventually made it West in his late teens. But in the end, he wasn't able to leave the industrial world behind. By the 1980s, from trails coursing up the western slopes of Colorado's magnificent San Juan Mountains, Gary watched as the air grew increasingly grainy—polluted in large part by power plants hundreds of miles away. The following year, out on the trail in the Sangre de Christo range of New Mexico, a doctor friend advised him to start wearing long sleeves; a thinning ozone layer, he explained, was increasing something called ultraviolet B radiation, which was in turn increasing risk of skin cancer.

By the 1990s, the land was showing the harsh side effects of a warming, drying climate. Since then, the problem has only gotten worse as industry and transportation continue pumping greenhouse gasses into the air. Big atmospheric changes in CO_2 levels continue to affect streams in Yellowstone, drying them out and devastating native trout. As a result, the eagles that feed on those trout can go hungry, which can mean less success hatching and raising eagle chicks. Meanwhile, the bears do without the early summer flush of protein they once got from those fish. Some move on to make meals from plants in nearby meadows. Others venture out of the park to raid garbage cans in local communities. Some of those bears get shot.

From childhood forward, Gary had imagined those wild places as separate, protected from the abuses going on in more settled and industrialized places like where he grew up. Nature was good and cities were bad. Nature was safe tucked off by itself, far away from the excesses of industry.

Gary's love of the wild took on its own separation-based cynicism. His isolated nook was one of an environmentalist less focused on healing what was wounded than on preserving what seemed whole. Fortunately, nature is a good teacher. Ever so slowly his illusion dissolved. The more time he spent in the wild, the more he came not only to admit, but also to take comfort in knowing that nothing was separate from anything else. The urban world of his childhood was never separate from the wilds. Cities thrive when nature as a whole thrives.

And conversely, when nature suffers, we suffer with it.

Again, separate thinking can have an important place. You isolate and compare things all the time, employing separation as a tool of discernment. What's more, there've been great benefits from the particular kind of separate thinking originally known as subject-object science; over the centuries it's led to everything from x-rays to airplanes to solar panels.

Yet this way of thinking took Western culture like a tidal wave. Objectification went from being a clever tool of scientific exploration to a means—one woefully incomplete—of knowing ourselves. Our certainty about the reality of separation is what keeps reinforcing the consumption-driven, techno-mediated, hurry-up-and-perform arenas in which we live. And yes, each of us contributes to the problem every day.

How is it possible to have misplaced our highly evolved instinct to connect? After all, you probably think of and actually experience the natural world as a haven. So, what keeps the comfort of that haven from pervading the whole of your life?

First of all, as we've established, separate thinking sits at the very core of Western cultures and economic systems. "You're on your own." "Prove yourself." "Pull yourself up by your own bootstraps." As a matter of pride, we routinely praise the idea of the rugged individual, even though there's no such thing.

Quite young, you learned to think about and act in the world in ways that confirmed the "truth" of your separation. Even a quick look at recorded history teaches that nature is primarily a resource to be exploited by clever and enterprising captains of industry. Forests are for houses. Mountains are for drywall and jewelry. Some animals are for lunch and dinner, others to greet us when we walk in the door. To a degree, none of this is bad. But when our engagement with other living beings is shaped only by the story that humans are the owners of life, things go off the rails. A great many people might respond, *That's just the way it is.* But it's not.

Then there's everyday conversation in the office, your living room, or at the market. You're swept into "binaries" of conservative versus liberal, smart versus dumb, emotional versus stoic, rich versus poor, black or brown versus white, straight versus gay. All around you, all your life, you've watched as people pick their teams—teams formed on the basis of shared traits, those traits too often used to signify worth. Once again, you define yourself not just by the people on the team you've chosen, but also by the people on the other team: your opponents. Without even knowing it, you can end up living your days isolated in a tiny corner; the astonishing creative power of your interdependence stays out of sight, out of reach.

You wouldn't be sitting with this book if all of this weren't wearing thin for you. The ground of exclusive separate thinking is a flimsy and vulnerable place on which to stand. And quite likely, no one in your everyday life is saying anything about it.

So, you stop. The more you watch for separate thinking, the more you find it—unmasked in the headlines, at the ballpark, at the doctor's office. Then you stop again, this time long enough to catch yourself in a full-on *me story*. *Those damned politicians; they're driving me crazy! All I want is a quart of milk; I can't believe I have to wait in this line.* Every separate thought requires a *me* and an *other*. It may seem like this level of scrutiny could leave you overwhelmed, unable to move. But we're only asking you to *notice*. Because in that space of just noticing, without any soundtracks of judgment, more connected ways of seeing will naturally begin to grow.

By this point separate thinking isn't just entrenched—it's hungry. And one of its favorite feeding grounds is evaluation. *I hate the way I look in this dress! Millennials are so lazy! Boomers are clueless! (Fill in a party affiliation) are idiots! Here I am walking to work, while all these other morons are driving. It's hardly worth us worrying about climate change until China does its part.* Such evaluation weirdly bolsters the ego. It allows you to hold yourself in a favorable light based on the less than favorable status of others. And being better than someone is—you guessed it—a sure-fire way to feed separate thinking.

But it's all empty calories: a twisted game of subject-object thinking, where we get to run around playing the subject, making pronouncements about inferior objects. We indulge ourselves in acting as if the world is made up of stand-alone life forms that are either useful, of no concern, or potentially threatening. Increasingly, reality is what we claim it is. The way *you* know the world is the standard, and everyone else needs to shape up. As an arbiter of evaluation, you get pinged just enough with moments of feeling smart, or vindicated, or just plain right, to keep the game going.

From age nineteen on, Mary had a mentor. Mayme Porter, she says, was her "fairy godmother." Mayme was raised in the parched plains of West Texas, born just as the Dust Bowl kicked into high gear. She'd tell of sitting as a child on the edge of a pickup bed, the threshing vent coughing dusty grain at her feet, slowly piling up to her knees and thighs. Her job was to use her small fingers to remove the grasshopper and cricket bodies from the harvest.

She never said much in particular about her family or community. But one afternoon under the shade of a mesquite tree, she did say, "We teach the way we were taught. We parent the way we were parented." She and Mary were sitting there in the dappled summer light, each holding a sweaty glass of iced tea. "That is," Mayme continued, "unless we have an exceptional experience, someone or something that shows us another way."

It wasn't until years into her career that Mary realized what made Mayme's words so helpful. How with everything she spoke about, Mayme used *description* instead of *evaluation*. The benefits of that way of speaking came into even sharper focus when Mary had the opportunity to work closely with the elected and traditional leaders of the Tribes of the Pacific Northwest.

Across twelve years, she took direction from a group of tribal leaders working to increase the enrollment and graduation of Native American students in college and graduate school. Among the many things she learned was to listen better. She slowed her impulses to speak and began to hear how the Elders, like Mayme, consistently used description. This was true even when they were speaking of extremely difficult things: Rising suicide rates. Rampant diabetes. The presence of non-Indian meth labs on reservation lands. The terrifying frequency with which Native American women are missing and murdered.

As Mary quieted herself to listen and learn, she saw how easily her white defensiveness was activated—how separation thoughts like guilt and overwhelm came rushing in. Fortunately, she'd gotten better at listening past those tired habits. And when she did, she could hear, with every story of early European contact, every story of the Great Death from smallpox and other diseases, every story of racism still alive and well, the truths the Elders revealed by speaking simply and deliberately, from description.

It went something like this. An Elder might speak about the shortcomings of education for tribal youth. She might place her comments in the context of the federally funded boarding schools that began in the mid-1800s—a system that took Native American children from their parents and their homelands to schools where they were punished for speaking their languages, made to dress in the ways of white people, forced en masse to learn and act with rules inspired by the military. Many were physically and sexually abused. As she spoke, tears might come to her eyes. But when she came to the end of her statement—to her call for quick, thoughtful improvements in the education of tribal youth—the Elder would not blame, nor be hostile toward, the non-Indians in the room. Rather she'd look out at all the people gathered. Native and non-Native. And she would say, "This is the situation before us now. What do we do together to make it right?"

The Elder's question calls forward our best descriptions of the circumstances at hand. It calls for another shift in attention. As Mary began telling Gary about her experiences with Mayme and later with tribal leaders, the two of us began to realize that these teachers had all been speaking from the fundamental interdependence of life. Each spoke of human circumstances—parenting, schooling, health, community safety and care—in terms of a large web of relationship. Their perspectives were consistently grounded in present-time, essential connection. *This is what is happening. This is what we are doing. This is what doesn't work. These are the gifts that sustain us. Given all of this, what do we do together to make it right?*

As counterintuitive as it can seem, stopping to describe what's going on, including what's going on inside us, always clears the path forward because description comes from attention to relationship. Evaluation often has the opposite effect, because it originates in separation.

Consider this contrast:

"That guy just cut me off in traffic!"

Evaluation: "What an ass!" (Backstory: the world is so full of jerks! If only people were more like me.) Action: ride the "idiot's" bumper while pounding on the steering wheel.

Description: "Woah, that was close! Scared the crap out of me."

(Backstory: (a) I really need to focus, or (b) It's lucky as hell that I was paying attention.) Action: breathe; slow down and get it together so I don't cause an accident myself.

No doubt about it: it takes effort to catch yourself popping into evaluation. Sometimes you play the role of judge, and other times you're certain you're the one being judged. Either way, you're in evaluation mode. That perspective only amplifies already well-entrenched feelings of isolation—feelings, by the way, that are routinely encouraged by people who stand to gain from your negative evaluations of yourself or others. Indeed, for a very long time, isolation has been a powerful tool for maintaining the status quo. Nefarious leaders have long known that when you're flustered, angry, and distracted, upset with "the other," your energy and vision are compromised. In that state, it's hard to find the calm needed for steadfast action to effect real change.

So, we say stop. Practice description. For now, when you reflexively go to evaluation, just catch it. Describe what's happening, whether it's inside you or outside. Resist indulging in either justification or self-criticism. When you're less antagonistic and defended, you're far less likely to get stuck in evaluation, and that frees your attention. Your ability to simply describe helps you clear the fog to reveal how you've been seeing the world.

The two of us spend a fair amount of time watching for our own unconscious participation in the separation myth. Not a week goes by when there aren't plenty of examples big and small. Not long ago, into the first spring in our new home, our separate thinking got activated in an embarrassing way. The culprit? Well, to be honest, it was nothing more than the weather. We live in mountains, which can mean stubborn, lingering winters. But on the far side of that first winter, as the official onset of spring came and went, all we got were days and days, week after week, of overcast, windy, cold, wet weather. We didn't handle it well. We woke in the mornings feeling heavy. Cranky. Put upon. As petty as it sounds in hindsight, taking the weather personally grew in us like a fungus.

"This is an arid ecosystem, not a rainforest!" Gary pouted.

"How are we supposed to stay the least bit lively in all this darkness?" Mary pleaded. "And after that brutal winter!" one of us always added on cue.

And right there, in that small, unremarkable whine fest was our default to disconnection, to evaluation and separation and *me, me, me*.

Things turned, finally, when we surrendered yet again to the fact that we have not an ounce of power over the weather. With that realization, we remembered that we *did* have the option to take a break. We cleared local obligations and went mobile for three days—working from the road. Consulting weather apps for the nearest sunshine, we drove west and ended up on a stretch of BLM prairie in the middle of the Columbia Plateau. A windswept place rather rudely named "the scablands."

Every time we'd been anywhere near this landscape before, we'd seen nothing but brittle, dusty, dried-to-brown expanses coughing under a lackluster sky. That spring day, though, the prairie was anything but scabby. It was a miracle. The rain had just stopped and, as we pulled into the deserted parking lot, the sun started making a few cameo appearances. We got out of the car, stepped onto the trail just beyond a livestock gate, and started walking.

Almost immediately, what we thought we knew about these western grasslands totally evaporated. Our habit of dismissing them, based on a few high-speed road trips along their outer edges, was badly informed. We soaked up every next surprise. The flourish of vivid purple lupine larger than any we'd ever seen. Wild iris nodding on thin and limber stems—their pale lavender blooms strung like beads across the edges of the meadows. In places of bare rock and sage were exquisite pink bursts of bitterroot. The path led to bluffs above lakes resting in cups of rock, dotted with clusters of towering ponderosa pine. In the largest of those lakes were pond lilies in quiet bays where pelicans rested. Raptors circled in the sky, and along the shore were footprints of mink and bobcat and coyote.

For all our whining back in Montana, the truth was that if the weather hadn't sucked, we never would've found this hidden paradise. This land we'd written off as drab and uninteresting was in fact dramatically beautiful. As with so many other aspects of life, spring is reliable. But it comes when it comes. Our weather-weary complaints gave us an opportunity to get real, to go descriptive. Without that, we would've stayed cozy in our stories of being victims of circumstance. In the end we accepted the dismal weather for what it was, letting it push us out into a landscape that surprised and deeply delighted us.

Neuroscientist Anil Seth reminds us that we don't just passively take in the world; we actively generate it. "The world we experience comes as much, if not more, from the inside out as from the outside in." In other words, how we think determines what we see. Meeting life with description offers us a wider field of vision—a more open landscape, free of the compulsion to make false choices based on knee-jerk evaluations.

Description primes the pump for seeing a richer, wider world. Instead of resisting, we dare to befriend, or at least are willing to collaborate with circumstances such as they are. We start in relationship, not in separation. It can be helpful to approach climate change this way. For all the anxiety the problem sparks, we can challenge ourselves to stop, take a breath, and begin to shift our energy from feeding fear to fueling curiosity and connection.

Stopping is about seeing your habits and then choosing whether or not to engage them. You keep an eye out—with your friends, in meetings at work, in testy conversations with your relatives. Even in the middle of the night, when you wake up with your thoughts in a whirlwind. Though you might not like the facts of your circumstances, you can choose to stay descriptive. From there you can investigate your best options, and then chose an action, if there's anything to be done at all.

Here's something cool to think about: As you practice this kind of attention, new neural connections develop and strengthen in your brain. That means the power of description will become increasingly available as your first response. Your perspective will expand. Your world will grow bigger, and at the same time, you will become more unruffled.

To grow your description skills, you can simply note your surroundings as you go through the day—in particular, aspects of the natural world. We say this because most of us are already inclined to witness nature without nearly the amount of evaluative urge we bring to our interactions with people. This may partly explain why stress hormones in your brain actually diminish whenever you're out in nature. Shifting your attention to the natural world can leave you less mired in evaluative thoughts, inclined instead to notice the feel of the air on your skin, or the sound of mourning doves, or the smell of cut grass. Gaze into the beauty of a cluster of white clouds drifting across a summer sky. See if you can let yourself fall into one of those sweet moments of wonder. How would you *describe* it?

Notice the feelings that arise. The thoughts. What do you sense in the moments just after a rainstorm? How would you describe the feel of your footsteps falling on grass? Taking time to be intentionally aware of natural surroundings, even a tree-lined city street, calms your thinking and opens new ways of gathering information. As it turns out, your descriptive skills are well served by your senses, and the better you become at describing, the more fully you'll recover your links to the larger world. Paying attention in this way is an exhilarating alternative to going through your days barely knowing where you are. Taking time simply to describe what you see and feel in the natural world builds neurological capacities you can bring back for navigating your interpersonal world. And that contributes to the repair and well-being of all.

So, to sum up: We all default daily to the illusion that we're separate beings in a world filled with separate beings. Even when it comes to climate change, we can see our task as a matter of saving humans, rather than saving us all. The barriers to preventing and addressing climate decline reside in how we think, which is a big part of what led to the crisis in the first place. Finally, the most powerful way forward for humans is first to stop. And in the stopping, to quiet, to unveil the story we tell about life, and to see whether it reflects the reality of our nature.

All of this matters. But there's one more thing. If we're to move forward in the face of climate change—a challenge we created—if we wish to give our energy to relief and repair, we must grieve.

When we stop, grief will surely arise. That's actually a good thing. Grief is an essential way for humans to tap into our belonging. Grief is an inevitability of being in a life. But most of us don't really have much in the way of skills for engaging it. Even with climate breakdown howling at the window, we still tend to hold grief at arm's length. From the standpoint of separation, of believing we live our lives all alone, we too easily become convinced that grief is something we can't handle.

This is not working well for anyone, especially the young among us. The troubling rise in suicide rates for teens and young adults unmasks the failure of our social ecologies when it comes to living through grief and loss. If grief is thwarted, a sense of belonging, and all the comfort that affords, is less available. And because of that, a young person can interpret a disruption early in life as a reason to end it all, when it's actually a chance to learn how to be with grief.

Stopping to allow grief, painful as it may feel, pulls us into relational reality. It brings us out of isolation and puts us back in healthy interdependence, healthy community. But we need to show it, live it, talk about it, so that grief is a normalized part of our ecologies.

Humans aren't alone in living with grief and loss. Healthy grieving is shared by many animals, from wolves to magpies, sea lions to geese to dolphins to whales. When a death occurs among elephants, the cows gather underbrush to cover the body, recognizing in some important way the passage of a kindred life. Some herds then revisit the site, returning year after year.

Grieving can even happen across species. Mary remembers visiting her friend Merle on the Blackfeet Indian Reservation in Montana. While she was there, a beloved tribal Elder died. Her name was Molly Kicking Woman, a grandmother who across more than fifty years had taught tribal tradition to Merle. The morning after word came of this grandmother's death, Mary walked with Merle out into a sweeping pasture. Six horses were gathered shoulder to shoulder on a nearby hill, all facing east, all holding perfectly still. Merle leaned in to whisper how he'd never seen the horses do that before. "They're feeling the grief of the people."

For Gary, the messengers of cross-species grief were chimpanzees. He'd been asked to write a book called *Opening Doors*—a portrait of Carole Noon, who over decades, and with astonishing grit and ferocity, managed to rescue hundreds of chimps suffering in research labs. Against all odds, Carole was able to give them healthy, dignified lives in a beautiful Florida sanctuary. There were trees, loads of companions, and grass beneath their feet. Carole spent endless hours with each of those chimps—talking and laughing with them, offering comfort. On the evening Carole died from cancer in her home beside the sanctuary, the chimps—sensing her passing—emerged from their shelter buildings to stand together, silent. Solemnly, they faced Carole's house. Grieving together.

Life comes and goes. There's loss. It's a major part of the story you're living right now. By being born and growing your capacity for self-reflection, grief is guaranteed. Climate breakdown is forcing us to stand face-to-face with loss and death. And part of that encounter is laying bare the detritus of our mistakes. When you really stop, you're almost sure to notice a big, dark feeling of regret, as well as a longing for do-overs. Your first response might be to get busy, to separate, to just ignore it. But to allow grief, upon which our health depends, is to thrive in relationship.

Joanna Macy, an environmental activist and spiritual leader who started out protesting nuclear waste back in the 1970s, emphasized the profound limitation of any environmental effort when the people involved fail to admit their despair. In 1979 Macy wrote,

> The energy expended in pushing down despair
> is diverted from more creative uses, depleting
> resilience and imagination needed for fresh

visions and strategies. Furthermore, the fear of
despair can erect an invisible screen, selectively
filtering out anxiety-provoking data. Since
organisms require feedback in order to adapt
and survive, such evasion is suicidal.

When we fail to admit our despair, when we shove it down, we're in fact wasting our energy. And we're lying. Macy urges us to engage the experiment of holding and tending our grief while at the same time acting decisively from our best intelligence and strongest heart. That kind of action opens us to a bigger world, one large enough to carry grief and despair without being frozen by them. "See what happens," Macy suggests. "You'll be amazed how much more gets done when you stop pretending."

ASk

In this moment on planet Earth, it's time to face some basic questions: *Who are you really? What are the relationships that sustain you and help you know who you are—those without which you simply wouldn't be?* And just as important: *What's getting in the way of your natural, highly evolved inclinations to care for those relationships?* We understand that at first glance this line of questioning may seem way too self-focused, another round of navel-gazing. But rest assured: The doorways to truth that can be found inside you are the same ones that will lead you back out into the world again, better able to recognize and support that which sustains us all.

This act of asking—of exploring the social aspects of your daily experience—is the best way to assure that your anxious impulses to act or to withdraw don't derail your good, heartfelt intentions. Research in problem-solving shows that reactive responses end up addressing the wrong questions about 95 percent of the time, because too little attention was given to understanding the problem in the first place.

As against the grain as it can feel, stopping to get clear opens the way to real solutions. It uncovers the trance of separate thinking and makes possible the necessary work of investigating the difference between *being* separate and *having* the experience of separation. To ask, then, is to find out what keeps those separate beliefs so activated, so captivating. Asking that arises from a full stop moves you toward real answers.

––––––––––––––––

You've likely experienced the way asking questions can become an exercise in chasing your tail—one question leading to another, then to the next and the next. But it doesn't have to be that way. Anthropology, the study of human culture, offers an approach to inquiry known as *participant observation*. Participant observation is a tool of description. It happens where you are—in what's going on right now. It gives you direct access to catching yourself making moves—big or small—that either help or cause harm. With a simple shift of attention, you can be a participant and an observer at the same time—fully in your life, even as you're watching it unfold. Participant observation is self-reflection. It enhances your progress toward acts that really make a difference: a way of asking that synchronizes your life with the realities of relationship.

When you were little, there was a time when you'd had no experience with hot stoves. One day, curious about the whistle coming from a kettle, you reached to touch it—only to be swatted away, scolded. That was confusing. There you were in the middle of your natural curiosity and—*whap!*—but it worked. Even without the adult's breathless explanation you'd probably have made the connection. Hot stove = potential danger. Be careful, pay attention. You participated and observed the results. Fire, water, container, your reach, an adult's quick response—done. What you knew about the world shifted to accommodate this experience, to see a connection between fire and danger that continues to serve you well.

But most learning isn't so quick. It's incremental. Tennis strokes improve with small gains. Same with the mastery of a second language. Or proficiency playing the cello. Along the way, there are mistakes and there are improvements; you benefit from both by paying close attention. Participant observation lets you withstand the flush of humility that comes with seeing your mistakes, using them instead as signals for course correction. You come back to knowing that, like nature itself, you're at your most capable when your ability to self-correct moment by moment is fully engaged.

Without a doubt, a major cost of separate thinking is that it limits your ability to self-correct. Gary saw this up close while writing his book *Land on Fire*, in part through hundreds of conversations he had with scientists, historians, and front-line wildfire fighters. Wildfire speaks to a separation problem that started more than a century ago. A time when fire policies held some big, unexamined assumptions about what happens when a forest burns.

Back in 1910, a major wildfire known as the Big Burn occurred in the western United States. It roared through more than four thousand square miles in Montana, Idaho, and eastern Washington. At least eighty-five people were killed. At the time, there was a fledgling agency known as the National Forest Service, eager to flex its muscles, in part to prove its worth to the Congress who held the purse strings. The Big Burn provided that chance. In its immediate aftermath, the Forest Service began selling itself as the agency best suited to be the first line of defense against wildfire. Fire was not only dangerous to people, as the Big Burn amply proved, but it was also devouring national forest resources that a growing country depended on. To say the Forest Service was enthused about this mission is an understatement. In fact, not long after the Big Burn the agency boldly instituted a so-called 10:00 a.m. rule, a remarkable and utterly unrealistic show of confidence that called for putting out any wildfire by 10:00 the morning after ignition. Though the policy was naive, the larger public saw it as honorable—undeniably heroic.

Wildfire came to be treated as a public enemy: an evil force hellbent on destruction. For roughly the next seventy years, we put out every wildfire we could get our shovels on. Smoky Bear came onstage, soon to become the most widely recognized public service icon in the world, which he is to this day. But there was a problem. We never stopped to investigate the overall impact of fire in the ecosystem, both the good and the bad. Had we slowed down, had we asked some questions, we would've noticed that healthy wildfires (fires of lower intensity than the Big Burn) were cleaning the forest of its natural accumulation of dead and fallen trees, limbs, leaves, and needles. We would've come much earlier to understand that in the arid West, fire is the *only* way the biomass of that old wood gets released to feed nutrients back into the soil. This kind of recycling is critical to sustaining forests over the long haul. Without fire, that debris on the forest floor grew ever deeper, year after year, creating staggeringly incendiary fuel loads. In fact, today, all these years later, some three hundred million acres in the US—an area about three times the size of California—still have excessive fuel loads.

But today there's an extra twist, which is the heat, drought, and dwindling snowpacks driven by human-caused climate change. The combination of old fuel loads and a warming, drying climate has brought a startling increase in unstoppable so-called megafires. These are fast-moving infernos that roar through the woods with incredible intensity, burning one hundred thousand acres or more. Sometimes these fires are so hot they literally vaporize plants, which in turn coats soil with a silicone-like, water-repellent sheen, a barrier that makes it hard for the plants to regenerate. What's more, these hydrophobic soils just can't absorb rainfall, which means that not long after the flames go out, landslides and damaging floods are almost sure to follow.

When those land managers from the early twentieth century decided to fight fire at every turn, they meant well. But they moved too fast, and the actions they took came largely from unconscious, habitual thinking rooted in subject-object separation. Fire didn't strike them as a natural part of healthy forests, but merely as a force of destruction to be extinguished. It was one of those parts of nature, among so many, that they felt duty-bound to control. At the same time, the loss of trees to wildfire was understood as an economic loss to humans. Few took the time to see the damage happening to the ecosystems. If there'd been an inclination to stop and ask, to look again and self-correct, that pause would've brought forward far sooner the fire-management practices we know today. Now we allow wildfire to burn where it's safe to do so, in some cases lighting controlled intentional fires to clear the forest floor.

The decision to subdue even healthy wildfires stayed in place year after year through a lack of attention to the larger world, to context; we ignored the fact of interdependence. Leading with relationship, growing our ecological intelligence, helps us see and then correct our mistakes more quickly. Think of it as a healthy leap from thinking of ourselves as *resource managers* to developing our role as *stewards*. Compelled by our interdependence with living things like forests, we become participant observers, humble enough to seek guidance from the natural world.

It's not news that dominant Western cultures developed to be downright intoxicated with control and certainty—with the fruits of presumed power over nature. It didn't take long to get so giddy with what we did manage to control, that we started thinking we could control *all* of it. With that, the sensibilities of "civilization" planted themselves outside the realities of life itself. In the ensuing centuries, individuals and institutions have played along—willing to be wooed over and over again with the illusion that we're exceptional, that we're in charge. We don't seek answers from nature when we think nature is more or less irrelevant. And that leads to problems.

We enhance crop growth with massive applications of chemicals, only to decimate bee populations. Fewer bees mean less propagation of the flowers that become pears and cherries and snap beans, tomatoes and peppers—many of the same crops we try to enhance. We drain aquifers to sustain sprawling urban centers, corporate farms, and industrial empires, thus deteriorating water quality and reducing the flows of streams and rivers.

Are we just slow learners? Why, after all this time, wouldn't the fact of our deep interdependence be the foundation for everything we do? Why aren't we more inclined to really ask what's going on in the world around us? Part of the explanation can be found by looking at the way humans, like all plants and animals, grow—the way *you* grew. A place to begin asking is right at the beginning, considering the way your ability to think as an individual developed in the first place.

Some of you are parents. Even if you aren't, you've likely been around babies enough to have noticed that time early in infancy when they really have no sense of being individuals. No sense of *me* and *you*, of *this* and *that*. For the infant there's simply no story. She's indistinct from everything. In her world, she's the one who cries *and* the one who comforts her. It never even occurs to her that she's somebody separate. But after only four to seven months, everything changes. This is when what developmental psychologists call *object permanence* begins.

Here's how it happens. Imagine a three-month-old in her recliner. She's all dry, recently fed, and smiling up at you with what looks to be pure uncut affection. Irresistible, right? So, you linger. You follow the urge to play with her, to start into one of those goofy grown-up tricks that infants are so good at eliciting. Maybe you make raspberries, pursing your lips and blowing so they vibrate. Her smile dances with tiny cascades of baby laughter. You're hooked!

To keep the joy going, you try something new. You pick up one of her brightly colored toys—say a rattle. When you shake it in front of her, she watches. Then you hide it, thinking that she'll get the joke; that she will show some sign of missing it and that, taking the role of beloved hero, you'll bring it back into sight. But when the rattle disappears from view, she glances about unfazed, as if it never existed. She shows no interest, no curiosity. Out of sight, out of mind. Because in her reality there's nothing except what's immediately in the moment. In the instant the rattle is no longer visible, it's no longer real.

But you're a die-hard. You keep trying out the rattle game from time to time, for months, until sure enough, one day you get a different reaction. Instead of her face going placid when you hide the toy, she frowns. She makes complaining sounds. She stretches to look for it. She's having the brand-new thought, "That rattle still exists, and I want to see it again." There it is: The first glimmer of a separate self. Self as distinct from all else. And just like that, the experience of being something separate from other things sets in for a lifetime. It's completely natural.

You too were an infant who went from having no idea of self, into the experience of knowing that things still exist even when they're not directly available to your senses or perceptions. You know the sun is still shining even on a day thick with clouds. You understand that the funds from your paycheck exist even when you've deposited them into a bank account. You know that giraffes walk the savannas, even if you've never actually seen a giraffe or a savanna.

This crucial transit into experiencing yourself as a separate individual is what led to your astonishing superpower—the capacity for self-reflection. Quite actually, coming to experience self as separate is essential to self-knowledge. It's the cornerstone for participant observation and self-correction. Thinking of yourself as a separate being pushed you to learn how to be with other people. You figured out the rules. You learned distinctions between good behaviors and bad ones. Rules made life easier—or at least made it make more sense. But by the time you were a teen, the rules weren't enough. You gained the ability to think more abstractly and could see that sometimes the rules don't apply. You also began developing functional ideas of what was fair and kind. You could understand the logic of consequences in a way that made the future interesting to ponder. Along the way, you also became able to imagine other people's experiences, which helped you build enduring relationships with friends and colleagues.

However, across a lifetime, the self-as-separate thinking that first took hold in infancy gets a lot of reinforcement. There's just something about being in a body—about the experience of *me separate from you*. It feels immediate and very real. And it is. Bodies are vulnerable things, and the fact of living *in* a separate body has provided every one of us with plenty to watch out for. So, it's easy to equate being in a separate body with being singular and isolated—on your own, like it or not.

It's important to ask into this. By promising more control, the separation myth seems to salve the vulnerability that comes with having a body. On the surface it can feel less complicated and safer to live as if we are only separate. We can limit our concerns to our separate bodies, our separate jobs and families and beliefs; using them as best we can to produce, survive, stand out.

But nature keeps calling, including the nature within. Not at all invested in whether we're listening, the natural processes of life nonetheless stand ready to remind us of the staggering degree of belonging, connection, and collaboration that sustains us all. This, then, is the paradox: To even have the chance of knowing interdependence, humans must first experience separation. By nature, separate thinking is a stepping-stone. And right now, the next step is vital.

———————————

Two primary forces keep the separation myth playing as the main feature at the theater, looping reruns across whole lifetimes. The first is the experience of being in a separate body—the quandary of cognitive development we just considered. The second is the limited lens through which we've been taught to see the world.

That lens came to dominate in the seventeenth century, put firmly into place by the authors of the Western scientific revolution. They were the ones who adopted a powerful tool for exploring the universe: a separation-based research strategy known as subject-object thinking. It called for the person doing the research (known as the subject) to completely isolate the thing that was being studied (the object). This was an act of separation on two levels. First, it required investigators (the subjects) to believe themselves fully removed from the world—a distancing that was necessary, they said, to ensure the purity of their research. At the same time, the object under study had to be isolated, observed as separate from everything else. They considered this a way of interrogating the world by rendering it less complex, less confusing.

If, back then, you wanted to figure out the essential nature of a frog, your efforts began and ended with the frog itself. Forget the qualities of the pond where the frog lived, or the range of insects it ate, or the egrets and kingfishers trying to feed on it. Looking in this way at just one frog, holding it in isolation, would allow you to make clean, unambiguous observations and measurements. Those facts, in turn, became the official story of what it was to be a frog. As the decades wore on, other researchers would drill down even more—looking at how the frog reproduced, trying to puzzle out the mechanics of its jump. Sometimes those investigations would lead to profound new questions; and those, in turn, gave rise to even more tightly focused fields of study.

By the end of the seventeenth century, subject-object investigations had come to be known as the scientific method—a cornerstone of modern science to this day. The idea of "disenchanting nature," as it was fondly called, was intended to make possible its prediction and control. How could we monitor and manage agriculture to ensure bumper crops? What machines could be concocted to mine precious metals? What chemicals could be applied to ward off hungry insects? The burgeoning capabilities offered by subject-object thinking proved heady, to say the least. Over time we used it to foster everything from weather forecasting to water pumps, airplanes to antibiotics. Indeed, the very technologies we're rushing to develop in the face of climate change come in large part from this kind of enterprise.

Separation science wasn't a mistake. But it's vastly incomplete. The problem—and this speaks directly to climate change issues—is that we've fallen so in love with separate thinking that we've written off other ways of knowing as being squishy or inconsequential. In the middle of the twentieth century, the brilliant behavioral scientist B. F. Skinner summed it up this way: "[Science] must restrict itself to what we can see and what can be manipulated and measured in the laboratory." While love, kindness, and empathy may exist, Skinner went on to say, those things can't objectively be measured. Which means they're not of interest.

All along, not one researcher, not even B. F. Skinner himself, ever operated in isolation. Skinner could elect not to trouble with studies of social nuance like kindness. But never was he nor any other researcher able to think or function without the relationships and conditions that made their work possible.

Think of it. Skinner's science was always socially situated. And the subject-object scientific tradition as a whole never bothered to admit that. Their investment was strong in the idea of objective truth—truth that could only be discerned through separation. While their findings were arguably rigorous, their conclusions were only ever a small part of the story. The problem, and it's a big one, came with their belief that objective analysis yields complete knowledge, keeping relationship always in the shadows. Full Ecology's practice of asking draws this serious omission into the light.

By early in the eighteenth century, science—and the separate thinking it championed—was the undisputed king of the Western world. Every area of culture would come to mimic it, privileging that which could be measured and predicted above all else. To this day, separation science remains a kind of Rosetta stone for developing mainstream education systems, economic systems, physical and mental health systems. As human society has rolled out in nation states and then into the corporate empires of today, it's carried in each iteration both the powers and the limited vision of subject-object thinking.

Dominion over the Earth, manifest destiny, and white supremacy are only possible as separation-based ideas. These worldviews operate by designating a few people as subjects with the power to control the objects—everyone and everything else. People in power have used subject-object sensibilities as seedbeds for devastating imperialism and colonization. The damaging effects of this heritage on the descendants of Native American tribes and enslaved black people still play out every day. In addition to justifying the oppression of millions, separate thinking is also at the root of deforestation, watershed destruction, and nearly unbridled oil and mineral extraction. The relational costs of these practices have, for centuries, gone unquestioned.

Again, Full Ecology calls for another way of exploring—of asking—beyond the traditional subject-object approach. In the early 1990s, Mary began asking about the unquestioned authority of standardized testing. Particularly in schools, significant decisions with big implications for learners were driven by standardized testing protocols. Diagnosis of disability, and related modifications to instruction, came from measurements forged by subject-object thinking. Learners were reduced to numbers—intelligence quotients (IQs), alongside quantities indicating everything from verbal, numeric, and spatial skill to emotional stability.

Mary wondered several things. Are these quantified pictures actually useful when it comes to supporting learning? Doesn't it make sense to start with data that are less abstract, less removed from the learners themselves? Her investigation led to this important finding: The best, most relevant data come from those closest to the learning itself. When assessment starts with listening to the actual experience, and thus expertise, of the person closest to the learning—the learner—it reveals better and more practical solutions. Then, including the observations of those most closely related to the learner's experience—teachers, parents, peers—makes solutions even clearer.

Sometimes the process bogs down. No matter the input from those closest to the learning, it can remain unclear how best to support the learner. At that point, quantitative measures can be great tools for getting the process unstuck—for revealing new questions to investigate within the relationships that support learning. With that, the usual tests can be used as what they've always been: tools and never the whole truth.

This expanded way of asking—of rooting our explorations in relationship—is as critical to climate change as it is to the classroom. Yes, we understand now that smokestack exhaust in Chicago can raise the temperature in Sydney. But fixing that problem doesn't start and end with technology. Instead, it begins with human relationship: What have we done to bring on this breakdown? How willing are we to come together in service of the common good? How able are we to understand the planet as a communal lifeline and to ground our initiatives there? The energy held in asking from a place of connection allows solutions to blossom.

The separation myth holds strong when we fail to ask in connected ways. It cuts us off, forcing us to live in tight and static realities. If climate repair is to be a possibility, we have to practice relational innovation—the kind that turns to the messy but thriving real world to uncover dynamic strategies for preserving the well-being of all life.

More relational investigation anchors the recently released United Nations *Global Assessment Report on Biodiversity and Ecosystem Services.* The UN profiles the worth of what human society considers the "free" services of the planet. These "natural resources" (a term fairly oozing with separate thinking), include clean water and air, as well as the pollinators that make possible many of the planet's food crops. Just the annual global value of crop pollination by bees, insects, and other animals is estimated to be a staggering $577 billion. Meanwhile, here in the US, all of nature's services taken together are at this point floating the nation's economy to the tune of more than $24 trillion a year. That's more than $73 thousand per person.

A gold mine on public land in California may "pencil out" to show a tidy profit in jobs and taxes; for a very long time that's exactly how such projects have been approved. But if that mine is located at the headwaters of a major stream, and ten years from now there's a cyanide leak from the leaching ponds, suddenly the cost of cleanup—not to mention paying for any associated health-care costs—can force the balance sheet into the red. Asking only from a base of separate thinking can lead us to make economic decisions that don't bother to consider the value of healthy, intact ecosystems.

So, let's make the asking personal. What history were you taught that you don't question? What worldviews have you taken on from your family, your spiritual tradition? How reflexively do you think of your body as an annoyance, separate from you? What about nature? How sure are you that you're totally disconnected from the pigeons convened on a rooftop, or the neighbor's cat rubbing against your legs? What mechanisms chug along beneath your radar to fabricate your perceptions, beliefs, and choices?

At least one of these mechanisms likely has to do with the giant, amorphous, and persistent idea of productivity. How much is your sense of self-worth tied to what you produce? What are you hoping for? What are the payoffs? Not that productivity is in essence a problem. In fact, it's completely natural to any viable ecology.

When Mount Saint Helens erupted in 1980, its scorching mudslides, called pyroclastic flows, ran like molten floodwaters down and across the alpine ecosystem. They scoured forests and brought high mountain lakes to boil. Gone were trees and grasses, and in the lake, there were no more trout. Ash blew hundreds of miles beyond south-east Washington state, lingering in the air for months as the caldera smoked. On Interstate 90, just west of Spokane, snowplows arrived daily to clear massive ash piles that drifted onto the road. But here's what happened next, in nature's good time.

Twelve years later, in 1992, Mary and her daughter, Sara, stood on the edge of one of those pyroclastic flows, talking with a ranger. They were looking at a lake sprawling below, surrounded by a new forest of healthy lodgepole and ponderosa pines—ranging up to fifteen feet tall. The ranger gestured with some excitement as she reported that just that week, and against all odds, trout had been found in the lake. This high spring-fed basin should never have been home to trout again. But there they were, right alongside brave trees that had managed to seed, sprout, and thrive as ever. This was a natural expression of self-organization, resilience, and growth. Each individual organism doing its part, resilient only by virtue of diverse, interwoven community. A whole. The lake and the forest were productive, to be sure—but never in isolation.

But we humans often fall into seeking productivity in isolation, interacting with a vastly limited number of variables. We produce and produce, giving little thought to the well-being of the system, either social or ecological, that sustains us. The more growth, the more accumulation, the better. In this way, productivity can turn cancerous.

When Gary was a young man in the industrial Midwest, he took up the somewhat dubious practice of hopping trains. Rolling west out of South Bend he often ended up in the highly polluted industrial complexes near Whiting and Gary, Indiana—dumping grounds for hundreds of tons of benzene, xylene, PCBs, and hydrochloric acid. The ground near the rail yards was honeycombed with ponds rimmed by rectangular-shaped earthen banks, each one slick with sludge. Toxic fires smoldered at the edges of the neighboring city dump, smelling of sulfur and scorched wire. Years later it was found that the chemicals had leaked from the landfills, finding their way into the waterways and aquifers of Lake and Porter counties. By the 1980s it would be one of the largest Superfund cleanup areas in the United States.

Late one night in the rail yards, Gary struck up a conversation with a guy in his fifties named Stan—stooped, unshaven, wearing a dark blue T-shirt ripped at the shoulder. He said he'd been fired from his job on the slag line; he was trying to get west, to Kansas, hoping to crash for a couple of months with his oldest daughter. At a pause in the conversation Gary looked around, then said something about how weird the place seemed—the crazy-colored sky, the nasty smell, the puddles of fuel oil. Stan flinched.

"To hell with that," he snarled, turning to walk away. "You don't get it, fella. Poison is progress."

Stan's reaction may sound extreme. It definitely did to Gary. But his basic assumption—that production must win out at any cost—has been around for a very long time. And all these years later, it still gets plenty of play.

So why not sacrifice whatever is needed? Progress depends on production. If I produce the right stuff, after all, I'll make enough money. People will like me. Maybe I'll even get famous. And who knows, my success just might make me the kind of person that tragedy passes by. Holding onto these thin, precious hopes, we've dug in, accepting the mind-numbing pace of everyday life, the mandate to produce.

Neuroscience tells us that *neurons that fire together, wire together.* As we've seen, when you think of yourself only as a distinct person, it's easy to feel vulnerable and afraid, separate and alone. Meanwhile, society keeps telling you that being productive will ensure your personal safety and worth. The neurons in your brain that fire when you feel vulnerable, or even in danger, can easily be activated at the same time that your neurons for seeking security light up. Firing together, over and over, the two get wired. No wonder we're hooked.

Your neurology, then, makes productivity seductive. Wired with security, insulated from the anxiety of uncertainty, being productive is calming. Beyond the joy of work well done, or even the pleasure of a paycheck, to produce is to *matter.* Still, regardless of your productivity, there seems to be an awful lot happening without your approval. The control you want so badly keeps slipping away.

As we set off to ask these personal questions for ourselves, we got some help from nature. In the natural world, sustainable productivity embraces both individual agency *and* relational care. Some have equated these two with product (agency) and process (care). In actual fact, taking care of each other is every bit as vital as producing things, though for a long time it's sadly carried far less weight. The evidence that we routinely value agency over care is everywhere. Compare the status of teachers with that of corporate CEOs. Or of stay-at-home parents with almost anyone who has a job outside the home. Think of the income difference between a social worker and a land developer.

Unlike of the products of individual agency, care's effects are hard to measure in any definitive way, and even harder to predict. B. F. Skinner would say that since care isn't measurable, it isn't really worth paying attention to. Care is messy. It is the intricate give and take of being in relationship. A small shift for one player affects the whole concert. Yet every sustainable system of production arises within a relational web of care. Quite certainly, both agency and care are essential to a fulfilled human life. But make no mistake: the possibility of individual agency depends on the relationships that sustain it.

In wolf packs, both females and males tend the young, rotating child-care responsibilities while the rest of the pack hunts for food. Often, the same alpha female wolf that gives birth also leads the pack to find prey and to launch the hunt. Among mountain gorillas, males not only fiercely defend their families from threats, but also help care for babies. And while grandmother elephants live and breathe care, they consistently call the shots for when and where the herd migrates, as well as for how it defends itself.

You know from your own experiences with family and friends that relationship offers vital ground and support for your agency in the world. As we can see in the natural world around us, when individual agency and community care are balanced, there's no problem. But in much of our culture these things aren't balanced. We're far more inclined to look for solutions to our problems—be they personal or societal—in individual heroics.

It's important to understand that this tendency to overprize individual action requires a fair amount of privilege. The same privilege that allows us to establish our own safety and comfort can also be used to buffer ourselves from the direct, and sometimes messy, experience of interdependence. Such privilege gives us the option to imagine our success as ours alone—to believe that our ability to thrive is a solo show, and with that, to renege on care. We can find ourselves disinclined to slow long enough to explore the relational circumstances that allow for truly wise, sustainable choices.

This orientation to overvaluing agency is sneaky business. Even if you consistently support environmental health and social justice, you're susceptible. Notice, for example, how easy it is to hurtle into lone-hero mode, even as you advocate for a world of deeper relationships. See how often you still comfort yourself with feelings of being exceptional, superior to other people because you're doing something—and they're not.

At the other polarity from outsized agency, but equally as privileged, is the seduction of withdrawal. Withdrawal is often a kind of paralysis that comes from worry, guilt, embarrassment, uncertainty, or overwhelm about what to do—all of which result from separate thinking. To take time out to be paralyzed is a privilege.

The people and ecologies who are most directly affected by the fallout of humanity's separate thinking—coastal Inuit forced to abandon their villages in the face of rising sea levels, farmers fleeing droughts in North Africa, commercial fishermen struggling to make a living in the North Sea—are always in collaborative motion. They're not thinking about relationship as some sort of goal; they're living it all the time.

They're being danced by relationship every day—with weather, with land, with other people of their community. By necessity they remain tuned in, responsive. Neither impulsive heroics nor paralysis will ease the immediate reality of their circumstances.

All of this said, privilege by itself, like productivity, is not the problem. Writing from a lifetime spent at the margins, poet and essayist Audre Lorde reminds us that we all have privilege. The privilege of knowing our beloveds, the privilege of a gentle breeze, a birdsong, the food on our dinnerplate. These privileges tend to spark gratitude. Gratitude that can open us up to meeting more of the world, to taking a closer look. From here privilege calls us to responsibility—to asking hard questions, and then garnering the courage to travel where the answers point.

From a leaf turning toward the sun, to a lion species that over tens of thousands of years learned to run fast because its prey ran slightly faster, life adjusts itself, becomes more resilient, through relationship. Nature, in other words, constantly inquires and adjusts. It does a lot of asking. And a lot of the questions it poses are about relationship. Learning from our connections guides us to see the reality of what's happening around us; to tell the truth about our needs and the needs of the larger community and to respond with care and agency as thoroughly and as efficiently as we can.

Once seen, the separation myth can never be unseen. Your awareness increases with asking. It moves separate thinking from being something you "just can't help," to being one way of knowing among many. *Having* the experience of separation, instead of unconsciously *being* it, gives you access to much more of your human nature, including aspects that will support you through all the years to come.

You'll breathe easier too. Opening to your interdependence with all of life will reveal how, to an astonishing degree, the world always has your back. That understanding will help you be more naturally productive, and more naturally engaged in the care of both yourself and your relationships.

By now, you've probably noticed that reclaiming balance and building ecological intelligence involve a certain amount of discomfort. There are no free passes, no buttons to push to zap you into eternal bliss. A life is a life: connected, flawed, vulnerable, and brilliant. Going forward takes trust—trust that gets fortified when you open to the natural world.

There's long been a joke in spiritual circles that the infant—the being with no sense of individuality—is the guru. And, as an unaware teacher, he fits the bill. An infant gaze alone can be irresistible. Unwavering, quiet, peaceful, deep. Simple connected presence in warmth and safety, in natural exploring and curiosity, in being one with everything. If only we could flip a switch and be there again.

Alas, there's no such switch. Instead, we feel fatigue—fatigue that arises from trying to live as if we're separate. Use that fatigue. Let it push you toward the belonging you seek. The immediacy of climate breakdown is the mandate: Walk right in. This is your home. Your family.

It's the air you breathe, the air of your ancestors and your descendants alike.

Climate change is a strong reminder that it's often the unpleasant circumstances of life that lead to healthier ways of knowing and being—ways we wouldn't likely find otherwise. In that sense, upheaval—and climate change is nothing if not a stupendous upheaval—carries with it a chance to evolve into something better, to step beyond the false structures we've been leaning on for so very long.

Renowned Harvard developmental psychologist Robert Kegan has a term for the times in life that shake you awake and demand you change: *disruption of the embeddedness environment.* Back when you were a baby, fresh into object permanence, you knew the small rubber ball still existed when it rolled out of sight behind the couch. At first, that change was pretty benign; but as a new way of knowing, it set in motion a disruption of colossal proportions. On one hand it gave you new ways of having fun—peekaboo, hide-and-seek. But it also yielded previously unknown anxiety. Suddenly you'd disintegrate into wailing and sobs when your parents left the house, because you now had full awareness that they still existed. The world was really different. You were vulnerable, not at all in charge, and you had to adapt.

Of course, that was just the beginning. As you grew from childhood into adolescence and on into adulthood, you lived through a succession of disruptions to your embeddedness environment. Taking your first steps. Going to your first day of school. Reading your first words. Feeling the first buzz of sexuality. Falling in love. Losing at love. Getting a job. Getting a promotion. Becoming a parent. Getting a divorce. Struggling with a health issue. Losing loved ones to death. Your life has actually been a parade of disruptions—big and small, some welcomed, some not. Each time, though, you moved.

And here you are, by now meeting disruption with the added feature of choice. With each upheaval of your adulthood, you could have chosen not to grow, but instead, you broke the frame through which you had previously viewed the world and got about the business of building a bigger one. You gave up one way of knowing the world to inhabit a more inclusive one. And notably, with each disruption you kept everything you'd known before, adding to it rather than replacing it.

Consider what happens in the wild. In disruption and its aftermath, it's never a bad idea to ask yourself the question: *What would nature do?*

Over and over organisms and ecosystems are disrupted; over and over they recover, transcend, thrive, and quite literally, evolve. The snake sheds its too-tight skin to grow bigger. A eucalyptus limb broken in a windstorm sprouts new branches from the trunk. A grizzly bear, used to gorging on whitebark pine nuts in the fall, can't find them because of a rampant pine bark beetle infestation—and so turns instead to eating yampa and yarrow roots from a nearby meadow. A pack of wolves find themselves captured in Canada and later introduced into Yellowstone National Park, where they face the immediate task of figuring out how to make a living in a totally unfamiliar landscape.

We humans tend to think a lot about disruptions, being especially fond of imagining ones that haven't even happened. When the goal is to avoid uncertainty, every disruption triggers red alerts. *Danger ahead! Gain control immediately!* But here and now on this wounded planet is a very real and urgent call. We must tell the truth—admit where we are, right in the middle of not-knowing. It's painful and baffling, and that's how it must be.

If you shut down in the face of that pain and bafflement, if you refuse the call to change and grow, you enter what developmental psychologists call *foreclosure*. Foreclosure obstructs potential breakthroughs by walling off unfamiliar ideas. The person determined to hold change at bay, stoking that avoidance with anything from drugs to conspiracy theories, risks foreclosure. Foreclosure slows or even stops natural maturation by saying no to disruptions—by refusing to be moved, to be changed.

All this talk of healthy disruption can sound awfully inconvenient; it's AFOG, one of our friends reminds us—another f#@%ing opportunity to grow. But in every disruption, you always have choices. You can choose courage and empathy. You can also choose to foreclose. Be assured that the neurons of your brain will accommodate whatever choice you make by strengthening the related wiring. These choices thus become a core aspect of your identity. Who are you really? A person separate, or a being integrated with the whole of life?

Many years ago, each of us on our own came across an ancient Vedic tradition of spiritual inquiry called *neti neti*—which translates into "not this, not that." In a nutshell, *neti neti* is a 2,500-year-old way of investigating the nature of the true self by questioning and disqualifying things we often mistake for ultimate truth. In other words, the practice resides in understanding that the most sacred part of the self can't be expressed in any singular idea or concept. *Am I my name? No. Am I my body? No. Am I my emotions? No. Am I my job? My home? My parents? No*—until after many rounds, and through a good measure of disruption, you end up on the edge of a mystery. The true self is always present, but at the same time it can never finally be described with words or understood with thoughts.

But here's the thing. The answer, *I'm not this, I'm not that*, really means I'm not *only* this, or *only* that. In Full Ecology, we found ourselves understanding *neti neti* with a slight twist. Instead of focusing on what the self is *not*, we wondered about expanding our sense of connection with the planet by turning attention to the infinite things we *are*. It turns out the ancients were on for this pivot, too, calling it *Tat tvam asi. I am that too.*

Am I this tree? Could well be. After all, it's giving me the oxygen I need to live. It's also releasing chemicals called phytoncides, which with every breath fortify my immune system and strengthen my heart. And what I breathe out, it breathes in. Trees thrive, I thrive.

Am I this patch of dirt? Well, the dirt is teeming with microorganisms that break down essential nutrients for the roots of the plants I eat. As it turns out, my belly is filled with healthy microbes—most of them coming to live with me after I was born, some of them from the soil that grew my food. And it's a good thing they did. Without them

I wouldn't be able to break down the food I eat, releasing the energy to keep my body and brain going.

What about these rocks? Am I these? Granite and sulfur-bearing stones began eroding into the lowlands and oceans two billion years ago, releasing essential trace minerals like zinc and copper. Single-celled life forms used those minerals to build the first proteins, which led to new, multicelled life forms. Which in the end made the frogs, the trees, and me. *Tat tvam asi.*

It's an empirical fact. You can never be defined by one singular idea or thing, but at the same time you're fully connected with everything on the planet.

It can be astonishing, and maybe a little unnerving, to learn you're not who and what you thought you were. Or at least not *just* that. But keep looking, listening, asking. How does it feel to know that you share more than half of your DNA with an apple tree—and for that matter, more than 90 percent with your cat! Or that you and the robin in your backyard are not only both made of the same chemical building blocks, but the cognitive parts of your respective brains—what you both use for things like long-term memory and problem solving—are wired in vastly similar ways. Or that just like you, every life form you encounter "runs on sun"—either using it directly, as plants do, or using it a little further up the chain, as creatures who eat the plants.

Can you really find a line where you end and the rest of the world begins?

ACT

It's early spring. On the bank of a westward-flowing stream in the foothills of California's Siskiyou Mountains, the seed from a redwood pushes through its blanket of ground cover. The air is cool and moist—the sprout bright green and tender. Its arrival is an act. A natural act. An inevitability of conditions being just right.

This could be the end of the story. More often than not these delicate beings don't make it far—squelched by cold or drought, trampled by a still-sleepy bear, munched by a hungry elk calf. But in this case, the sprout hangs on, soon branching into two tiny limbs smaller than matchsticks.

The story could end here too. But there's a chance it will go on.

Camped in this magnificent redwood forest, we're surrounded by success stories. Every one of these trees is constantly engaged in strong and elegant action. Action that for all of them started with a cone dropping, maybe two hundred feet or more, from the upper branches of the giants. Each cone is filled with seeds so small that it takes more than a hundred thousand to make a single pound. Of the seeds that are fertile—and only about 15 percent are—just the ones that land in damp but well-drained soil have a shot at germination. From then on, it's one action weaving into the next.

To fend off trouble from insects, each tree is wired with a genetic capacity to produce chemicals that will make it unappealing. And in response to carbon deficits early in its growth, each tree knows exactly what to do; how to open itself to the underground fungal networks that connect it to the rest of the grove, finding helpings of nutrients from other trees. Much of the action is on the molecular level. When a photon of sunlight reaches the tip of a needle, it is instantly conveyed to cellular energy centers. In short, each of these magnificent redwoods is supremely efficient—adjusting moment by moment through every day, across every season, to rainfall, to drought, to fog and wind and freezing and pests.

In full connection with the life systems around them, redwoods perceive what's needed in any given moment and respond, often with the help of other trees, always in the most streamlined way possible. For most of nature, this is just how it goes. But as we've already seen, for humans things tend to be a bit more complicated. Living out of our deepest nature, launching actions that will serve the greatest good, involves thinking. It involves stopping and asking how we are socialized to act inefficiently, even destructively. It reveals the limitations of separate thinking. It also shifts attention back to your most essential lifeline—the reality of your full interdependence. Thus, having stopped and asked, you come to *act*, ready now to meet the question, *How do I live from relationship?* That question now orients you as an agent of healing for yourself, your family, and all life on Earth.

It is, of course, an open question. Beyond all the prescriptions for how to use less or how to shrink your carbon footprint, to really *live* from connection results from how you see yourself in the world—from your experiences, and the stories you tell about those experiences. By stopping and asking, you reach out to embrace a broader, more realistic view of how life works; you gain a strong sense of your true human nature. Such perceptual shifts of thought establish the interdependent vision that will move you ahead on the path of climate repair. Now is the time to act.

———————————————

Both you and the redwood are interested in efficiency. Redwoods are efficient in the ways of redwoods. Humans are efficient when guided and focused by reflective thinking—watching, listening, pondering, gaining insight. By truest nature, all acts of life are oriented toward surviving and thriving—toward Full Ecology. For our species, outside of the need to run from a lion or jump out of the way of an oncoming bus, failing to be honestly reflective can lead to misguided actions that end up diminishing our natural ability to thrive. We can actually harm the climate of our own bodies. Fuming for too long about how your idiot brother voted can leave you knotted up, and therefore less efficient, losing track of details, forgetting to get the mail or missing a meeting. Fuming activates your adrenal system, which in turn makes significant changes in your brain and body. If your adrenals stay elevated for very long, your blood pressure and blood sugar levels go up, and your muscles weaken (all of which, by the way, is likely to fray the strands of your relationship with your brother even further).

The goal isn't to do away with life's irritations. That's not going to happen. But if you want to realize your best chance at a long, healthy life—let alone be an agent of positive change—your actions need to be guided by sound reflective thinking. You're naturally wired to do this, and like the rest of nature, to be efficient.

The trick is to act in concert with the actual reality of the present moment—to act in ways that are rooted in calm, honest assessment.

By now you know that life is full of emotional weather, and there are still days to come when you'll be right in the middle of a storm. Reflective thinking can be shelter from the storm—to stop and ask in order to shape your best next action; to do something without anxious investment in specific outcomes. And that's really important. Life, after all, is tangled and dynamic. Trying to hold things still, relying on formulas to guarantee outcomes, yanks you right out of the present moment. It leads to mistakes and frustration.

Think of a hawk flying in gusty wind. The grace she brings to landing on a tree branch comes from constant awareness of the wind—how fast and from what direction it's blowing—making a series of split-second adjustments to her body and wings. The hawk has no ability to foresee changes in the wind. What she can do, though, is to be present; to react in precise harmony with her circumstances.

You may remember the environmental slogan, "Think globally, act locally." It's a saying credited to Patrick Geddes, a Scottish biologist, geographer, sociologist, philanthropist, and pioneering town planner of the late 1800s and early 1900s. "Character…is attained only in course of adequate grasp and treatment of the whole environment," Geddes said. "In active sympathy with the…life of the place."

We see Geddes's proclamation as a call to step up our ecological intelligence. To think globally and act locally is to ground every action in relational thinking—thinking that begins in your own direct relationships. Here locally, what you do always matters. It's on this ground that you stand the best chance of bringing the greatest measure of health to the widest range of beings.

Every morning you rise, and every night you go to sleep wrapped in relationships. You have relationships with your family and coworkers, your neighbors and friends. Your health is entirely a matter of the relationship of the sun to your home planet, tilted in summer to soak up the heat that lifts the water from the oceans that later falls as rain. The rain in turn is in relationship with the plants, helping them grow the flowers that expose still another relationship, which is that of calling in the bees and flies to pollinate them. Some of those flowers, in turn, will produce things like raspberries and blueberries for you to toss into bowls of cereal or ice cream.

Each component does its part. But every one of those individual expressions is brought to full power because local actions nourish community. And when you start linking communities together, well, the creativity gets absolutely profound. At a planetary level, the overlapping circles of community in action actually maintain the balance of atmospheric gasses, acidity levels in the water and soil, and temperatures across the planet. Taken together, all of this makes it possible for life to carry on. Constant, efficient local action.

Back in the late 1980s, Gary was asked by a publisher in New York to compile a book of nature myths from around the world. These were to be bite-sized tales about the origins of nature and its wonders—from how rainbows were hung in the sky, to how birds got their songs, to an ancient story of friendship between a forest and a tiger. As part of the research, Gary talked to storytellers and listened to archival recordings by anthropologists. He also spent countless hours in university folklore collections, perusing well over a thousand tales from around the globe.

Along the way, something amazing happened. After several months of research, Gary realized that without fail, every story he came across seemed centered in one or more of three qualities that people through the ages thought essential for living well in the world.

The first of these qualities has to do with sustaining a relationship with community—not just among humans, but woven into all creation. It is through community that the characters in tales from around the world gain not just knowledge, but also the contentment that comes from belonging.

The second essential quality Gary found in these stories is beauty. Beauty, which we perceive so easily in the natural world, has a special power to move us quickly beyond the *me story* into a state of awe and delight. From there, as it happens, we tend to shift quite easily into deeper kinship with the larger world. The stories Gary found showed that while beauty is often ephemeral, it's also reliable. It's always here, for those with the eyes and heart to see. Opening ourselves to beauty is a great way to ease anxiety and strengthen intention.

The third quality revealed in these stories is mystery, the unknown. Characters learn to welcome the times when life is unfathomable; to welcome them with generosity and surrender, embracing the shimmer of all they cannot know. So many uncertainties will never be answered. Satisfaction, these stories promise, will come to those wise enough to live with the questions.

Years later, an Ojibwa Elder would sum up the real takeaway of the tales Gary found. "Our people's stories hold life's lessons," she told him. "Bad things always get worse when you forget the lessons."

The three qualities brought forth in nature tales have been guiding humans for thousands of years. They still have enormous power.

A decade after finishing the book on nature myths, Gary had the chance to spend several months in the wilds of southern Utah with a group of smart, though thoroughly beaten down teens, many of them depressed and drug-addicted. Having come to the end of hope, and for some, very nearly to the end of life, they found themselves in one of the nation's best, most compassionate wilderness therapy programs. There they spent eight weeks living out of backpacks, supported by an amazing group of field staff and therapists. They learned to rely on themselves, on each other, and on the wild nature that held them in every hour. They learned how to tell the truth, how to listen, and how to speak so they would be heard. The impact of the program was astonishing. The success rate in treating drug addiction alone was nearly three times better than traditional twenty-eight-day lockdown programs.

Gary followed a dozen of these young men and women for a year after they left the wilds. In fact, these youth so moved and inspired him that he continued to stay in touch with them, interviewing nine of the twelve again a full decade later. At each point, he asked the same question: *Why do you think wilderness therapy worked for you, when every other intervention didn't?*

Without fail, their responses consistently fell into one or more of four categories.

The most common answer was, "It's the first place I've ever been where what I did mattered," referring to the fact that their choices in the backcountry had clear consequences for the whole community.

The second most common observation was, "It's the first time I experienced something beautiful."

The third reply was, "It's the first time I ever felt spiritual," or "felt God," or "felt like I was a part of something bigger than just me."

Community. Beauty. Mystery. Just as the world's nature stories teach, the three qualities were right there, living on, shining brightly in these fully modern lives that were once so hopeless and full of pain.

The fourth theme that came up during Gary's conversations with the young people from that program was a kind of basket that held all the others. It had to do with what they described as gaining the ability—and the courage—to tell the truth. On one hand, staying safe, eating well, and sleeping warmly required a very honest relationship with the wilderness. Being angry with a line of dark clouds overhead wouldn't stop those clouds from letting loose their rain. The truth of the situation lay in seeing the clouds for what they were, understanding what might happen, and then, with the help of the group, preparing accordingly. Notably, the consequences of refusing that truth were always completely logical. The natural world, after all, holds no hidden agendas. Choose not to put up your shelter one night, and you might just wake up to rain in your face. Fail to work together to build a fire and cook dinner, and the whole group ends up famished and grumpy, eating and then cleaning up well after dark. Out in wild nature, in full relationship with peers and guides, the separation myth made no sense. The link was obvious: Acting from the truth of every moment meant more success. It brought peace, and it brought empowerment.

These youth were living the promise of Full Ecology. Having been steeped in the natural world, they became keen participant observers. They stopped and asked, and then acted based on the living realities around them, with integrity and in accordance with their deepest human nature. We turn to their authority, to acts of *truth, community, beauty, and mystery* as anchoring themes for the rest of this chapter.

When it comes to taking action in the face of climate change, these four qualities offer an elegant set of "search images"—guideposts into the ceremony of reclaiming your connection with all things. Think of it like this: One day you're asked to go out and collect berries for the evening meal. As it happens, you've had a good teacher in the food foraging department, who's taught you that in this particular place the white berries tend to be poisonous, and the blue berries are bitter. The tastiest, sweetest berries turn out to be red ones, growing on bushes at least four feet tall. With that search image, you can head off into the woods looking at and assessing not every plant, but rather watching for chest-high, red-berried bushes. Which makes the whole enterprise a lot easier.

So it is when you choose climate change actions according to what serves truth, or community, or beauty, or mystery. By keeping any one of these qualities in the front of your mind, good and appropriate actions become more apparent. Try it for yourself. Now is the time to practice. Constantly.

And remember, every time you take thoughtful action you're engaging in a kind of ceremony. Ceremony is always about anchoring, about grounding key truths. Ceremony makes the energy of those truths more available to daily life. To the extent you can see your actions on behalf of the climate as ceremony—honoring something you really believe in—it will help sustain you in the years ahead.

TRUTH

Let's start with acts of truth.

Penguins naturally penguin. Eels naturally eel. Zebras are untroubled by any choice other than to zebra. But for humans to human, thinking is involved. This can be a liability or can be one of our greatest assets. Thoughts, insights, and intentions rooted in relationship lead to actions that match. Acts of truth *walk your talk*.

Acts of truth hinge on telling the truth most locally. Inside yourself. As neuroscience shows, the world looks the way you see it. The biggest barrier to meaningful action, in other words, is the noise and distraction of a faulty belief system. Telling your truth, acting out of the reality of the current moment, is an act of kinship. An act of love.

Film producer Christine Arena points to the resilience that builds in climate scientists who find the courage to speak honestly about their own despair. "More scientists are bringing their emotions and hearts to the forefront of their work—getting bolder, more impassioned, more provocative, [and] this collective grief is making their outreach more effective." As these scientists fully own the realities of climate change, including their frustration with their own roles in it, their credibility—and therefore their effectiveness—is increasing.

When telling the truth is the first priority, thorny problems open up to solutions. For example, consider the practice of restorative justice—a powerful way for communities to address local criminal and civil offenses. Mary learned about this model of justice on a nonstop flight from Seattle to DC, when by chance she was seated next to Chief Justice Robert Yazzie of the Navajo Nation. The two had spent hours reading and dozing, but never speaking. Only as the plane began circling, awaiting clearance to land, did they introduce themselves to one another and start talking.

Chief Justice Yazzie spoke about the Navajo justice system—what he referred to as the *Navajo Peacemaking Model*. "The Navajo understanding of 'solidarity' is difficult to translate into English," he began. "It carries connotations that help the individual reconcile self with family, community, nature, and the cosmos—all reality. [But] most importantly, [Navajo peacemaking] restores good relations with self."

He went on to describe a justice system grounded in truth telling that supports human flourishing, in contrast to practices that only assign fault and level punishment.

The Navajo Peacemaking Model arises from stopping and asking. It's an approach to justice focused on relationships and consensus-building—on healing actions rooted in the wider community. Restorative justice involves the full participation of victims, offenders, relatives, and other members of the community as a whole. It relies on everyone bringing and telling their truth. And those truths lead to very specific actions, from how the offenders will make amends, to the practical steps the community will take to build support networks for preventing similar offenses.

"No person is above the other," Chief Justice Yazzie explained. "In a circle, there is no right or left, no beginning or end. Every person in the circle looks to the same center as the focus. The circle is a symbol of Navajo justice because it is perfect, unbroken. A simile of unity and oneness."

The inherent truth revealed in the Navajo Peacemaking Model is that people's lives aren't separate from each other. In contrast, conventional criminal justice systems put the crime exclusively *in* the identified offender. The rest of us remain separate, blameless, resting in the thought, *I don't have anything to do with that.* Restorative justice supports a social ecology rooted in the free flow of respect and compassion, which in turn becomes the ground from which practical paths to justice can arise.

Chief Justice Yazzie described it this way: "Responsibilities are fulfilled by living a better life, by making better decisions. And that better life, those better decisions, come from communicating needs and interests clearly, both with self and others."

When the pilot announced that the plane finally had clearance for landing in DC, Chief Justice Yazzie leaned over, offering one last observation. "Over the time we've been practicing peacemaking as the way for resolving disputes, we've seen another benefit. Because it's a tradition of our culture, the people are more likely to practice it naturally. On Navajo, peacemaking can solve problems before they even start."

Restorative justice is an act of truth. You're probably not a chief justice. Nor, for that matter, do you necessarily come from a culture that's made a practice of addressing offenses by looking first at avenues of repair. Yet restorative justice is relevant to all of us amid climate breakdown because every one of us, to some extent, is an offender—directly or by choosing passivity. At the very least, we are members of the larger community who face these restorative questions. What circumstances allowed this to happen? How are we as individuals and as a community sustaining these circumstances? What do we need to do differently to prevent future offenses?

Acts of truth in support of climate restoration and repair emerge from being clear with ourselves and each other about our contributions to the problem. It's time to admit that personal choices are directly linked with everything from wrecks like the Exxon Valdez to bad water in Flint, Michigan; from Alaska Native Elders displaced from their ancestral lands because of thawing permafrost, to coastal

Indonesians fleeing as their homes are consumed by rising oceans, erasing shorelines forever. In the end, acts of truth are restorative acts of justice.

Tell yourself the truth first. How do you contribute to the problem? And how are you already acting for climate restoration? Then, come together with others to speak honestly about what *is*. Telling the truth energizes acts that extend local recycling and composting, or persuade local governments to establish policies for near-term zero-carbon emissions.

Consider Greta Thunberg. As a young teen, she began her celebrated journey as a climate activist simply by encouraging her family to reduce their carbon footprint. Next, she decided to take more public action by showing up every Friday to stand alone in front of the Swedish Parliament. She held a hand-lettered sign, which read "School Strike for Climate." She was acting from the disquieting truth faced by her generation—the ones who will inherit the consequences of our environmental abuse. Her work is about the restoration of community-wide justice, which includes the integrity of taking responsibility for her own actions.

Over time, Thunberg has become respected, even famous for her tenacity. But she was never alone. There have been plenty of other committed and courageous young people taking actions of their own. Ridhima Pandey, an eleven-year-old from India, joined Thunberg and fourteen other young people from around the world to submit to the United Nations a formal complaint against five of the world's most destructive polluters. Kisha Erah Muaña, a twenty-three-year-old activist from the Philippines, stood with Thunberg in Spain to call on global leaders to take "robust and lasting action" against climate change. All of this inspired Angela Valenzuela of Santiago, Chile, who joined with Luisa Neubauer, a young German climate activist, to coordinate Fridays for Future, the international youth movement dedicated to stimulating action on the climate crisis.

Check these people out. And check out Autumn Peltier, an Anishinaabe girl of Wiikwemkoong First Nation in Canada, who at thirteen spoke out about the urgent need for water protection, often addressing officials charged with establishing water policy and practice. Check out Mari Copeny from Flint, Michigan. At eleven, she began working tirelessly to draw government attention to the water crisis that had persisted in her hometown since she was a young child. And Artemisa Xakriabá, who at nineteen made her way from the Brazilian Amazon to a climate strike in New York City. There she spoke forcefully about the increasing destruction of her family's homeland—highlighting the undeniable truth that, across the globe right now, indigenous people are overwhelmingly affected by climate change.

Recently we had the opportunity to speak with a group of teens in northeast Texas, and when it came to sharing their thoughts about climate they were very clear. "We're mad," said Julie. "The adults tell us our ideas are important, that what we say is important. But when we try to talk about climate change, no one pays attention. We don't have any power at all." Then Stacey added, "The adults know what to do. We want them to get busy and do it."

Stacey is right; there's much we adults can and must do. In some communities, adults have joined with local universities to measure health effects surrounding oil refineries and mines. Parents of elementary students in Missouri are pushing for a switch to electric school busses. In Oregon youth and adults have teamed up, going door to door to let neighbors know about the solar panel subsidies being offered by local energy providers.

For many of us, one of the most important acts of truth telling comes simply from being parents. A few months ago, at a climate stress conversation we hosted at our local library, a parent named Cynthia was in tears. "I'm thrilled every day with my daughter—just that she exists. And every day I'm terrified for her future." Cynthia isn't alone. Confusion and fury, hopelessness and urgency are rising fast in anyone who loves and cares about the world's children and youth.

In our conversations with parents, we often hear stories of people panicked into inaction, distracted by confusion and despair. Many parents are so burdened with anxiety that they find themselves struggling to be present with their kids. The themes are consistent, and they're clear.

"I feel crazy not knowing how to respond to Margo's sense of wonder," said Zoe, another parent. "I get so stuck in anger. But even more, I get stuck in this awful feeling like I've done the wrong thing bringing her into the world."

To say that millions of parents are overwhelmed with the realities of climate change is a vast understatement. But the simple, courageous act of truthful conversation is having one consistent outcome: the critical recovery of connection. Talking with our kids affords the chance for us to build relationships with them based on respect. Instead of trying to dodge a truth they're already fully aware of, we step up to help them process it. Meanwhile, talking with other parents offers a sense of relief, a healthy reassurance that our anxious, hopeless thoughts aren't ours alone.

We remember Leon talking about his daughter. "The minute Sophie was born, all wrinkled and perfect, I felt something I'd never felt before. It was like this warm mist, a tender feeling of connection, and through her a connection with every life that ever was. It was actually overwhelming, but there it was. I knew I'd do anything for her. And anything for the people and the world around her."

At which point the tone of the whole conversation shifted. The fear, fury, and desperation that had brought these parents together transformed. Attention moved quickly from stories of separation to stories of this wild sense of oneness.

"Yeah. I haven't told many people about that," said Zack. "That feeling of love, it swept me away—like it stretched out forever from right here (pointing to his belly)—it was so real. It may be the most real way I can think." Zack looked around the room and laughed. "Maybe we've all been swept awake."

Coming from that place of being swept awake, the parents' conversation continued seesawing back and forth between speaking their fears and spontaneously suggesting ways for moving ahead. There was talk of wanting to get better with holding despair while at the same time staying present.

Anna chimed in, "When I look at my children, I often feel so sorry. Right now, though, listening to all of you, I'm realizing that my apology is just words without some change in what I do. My two-year-old son doesn't even know what the words, 'I'm sorry,' really mean. He'll learn by watching what I do after I say the words."

Lena continued, "Maybe 'I'm sorry' is a reminder to care. And care feels like activism. One thing I think we should do right now is agree to help each other out."The group was all on for that, signing and passing around the contact list someone across the room had already started. And with that they were off and running.

During every such conversation, the love these parents feel for their children is reflected in the love that every other parent carries too. Experiencing such a deeply shared value is as powerful as it is clarifying, revealing meaningful action steps. The connection is also a great boon to people's sense of commitment. Suddenly you aren't just one parent recycling and walking to work, battling the feeling of being overwhelmed, but you're ten or twenty people—creating a local food-purchasing group, ride sharing or car sharing, attending city council meetings together to push for sustainable energy policies. By linking arms, such groups are actually mimicking how things are accomplished out in nature, where the whole world turns on alliances. Which is why to most parents, connection feels so natural, so very *right*.

Stop, feel, and describe the deep interconnection that parenthood reveals. Tell the truth. Support each other. Follow "I'm sorry" with real action to change. Act from care. And every week, every month, keep reaching out for connection, for courage, for help.

Well into one of these conversations, Mary commented on the miracle of anyone ever being born in the first place. From there Ron, a candidate for city commission, was struck by an idea.

"What if we took ten minutes every day to sit with our children and be in total awe that all of us won the stardust lottery?" he said. "Stardust combined to make you. Stardust combined to make me. Everything is a different version of the same stardust. All of it belongs."

To judge Ron's suggestion as fluff is a big mistake. For one thing, he's scientifically accurate. But beyond that, there's more truth and energy to be found in awe than in cynicism. In our culture, being cynical is often associated with being cool; but it's really the intellect playing separation games, finding cheap ways to reassure you that you're the clever one, that you're "above all that." Cynicism pushes aside wonder, and with great bluster demands to lead. But it has no vision, no humility, no curiosity. And so over and over, it lands us in the same dark corner of the same small room.

Songwriter Raffi Cavoukian has been writing and singing songs for young children for forty years, including the terrifically popular "Baby Beluga" and "Down by the Bay." Lots of the parents in today's climate-anxiety conversations grew up listening to Raffi. In recent months he's shown up again, asking if today's parents have enough joy in their own hearts to lift their children forward into the future. Reaching that joy, perhaps especially in the face of these troubling times for our environment, means walking a path that begins in humble truth. Climate repair starts up close, as locally as your next thought. And right action comes from that.

COMMUNITY

We will rise to the occasion of a challenge like climate change not by acts of lone heroism, but by the energy and surety of community. Loosely quoting Einstein: Genius is a team sport. Once again, this mirrors the timeless logic of nature. We know that forests recover from the disruption of wildfire by acting in concert to restore the entire web of life. Microbes in the soil set about their work making nutrients available, even as the latent seeds and surviving roots of grasses and forbs spring into action to stabilize the ground. Bees and flies and butterflies show up to provide pollination. Then, the first trees to recover, like the sun-loving lodgepole pine, make shade for the sprouts of slower-growing species like Engelmann spruce. Working always as a system, with each species influencing the others, wild nature shows an exquisite ability to garner immense strength through community. And, quite actually, so do we.

Of all our human behaviors, acts of community reflect the core genius of how life on Earth really works. Through community our actions multiply, strengthening the whole system. One person who knew this especially well was the mid-twentieth-century social activist Jane Jacobs. Working as a journalist in New York in the 1960s, she witnessed the crushing impacts of so-called urban renewal. At the core of those renewal initiatives was the practice of leveling entire neighborhoods to build multi-lane freeways, relocating the residents to isolated, out-of-the-way housing developments.

What makes Jane Jacobs so inspiring is the way she took on big developers and city government with arguments anchored in the natural world. And often, she won. Jacobs emphasized how the same diversity that made nature strong made human communities strong too. She pointed out that thriving urban communities drew energy and resilience through dynamic interaction—conversations on the front stoop, in shops, in restaurants, in dozens of small factory break rooms. She called these exchanges "the sidewalk ballet." Just as nature uses the interplay of community to constantly adjust, Jacobs pointed out that human culture remains vital by much the same means. She had seen elsewhere how gutting neighborhoods and moving the residents to housing projects was a decidedly "unnatural" act—one that shoved thousands of people into lives of separation and despair. Thus upended, the creative base of those neighborhoods disappeared, and local economies languished.

Out in nature, the consequences of destroying community can be seen even in the wildest of places. In the 1920s the last wolves were trapped and shot in Yellowstone National Park—actions taken by the National Park Service itself, which like many at the time thought wolves a threat to animals like elk and deer. With the wolves gone, the elk took to hanging out on the banks of the park's streams, eating sedge and mountain bluebells and geraniums and willow shoots. In fact, they ate so much that in some places all that vegetation disappeared. Without the roots of those plants, the banks crumbled. And other creatures who were feeding on that vegetation vanished.

Then in 1995, armed with a far better sense of ecology, the

National Park Service and US Fish and Wildlife Service reintroduced the wolves into Yellowstone. Almost immediately, the elk became wary, moving off those stream banks to places with better views and more running room. With the elk off the banks, the flowers and willow shoots gradually returned. And because of the willow, the beavers found happy homes in the streams again. The ponds the beavers built then made perfect habitat for yellowthroats and song sparrows to come back too.

Such is the irrepressible force of community. Every aspect of an ecosystem acts in concert with every other—adjusting, waxing and waning in the rhythms of life and death, emergence and return. Human nature has its own unique expressions of learning, of trial and error. Yet if Jane Jacobs was right—and we believe she was—solutions to a problem like climate change don't just flow from the top down, but also from the ground up, all of us acting together.

Acts of community have generated lots of really useful "to do" lists: recycle, compost, use LED lights, flush less often, buy local food, minimize your use of fossil fuels, shift to alternative energies, eat less meat, insulate your house, plant trees. Vote. Run for office. Support candidates who actively plan and act for climate protection and repair. Support organizations devoted to climate health. One such organization is Project Drawdown, which, by the way, has generated what we consider the most powerful and relevant list for climate action. The organization carries forward the work of Paul Hawken, who with colleagues across the globe identified the top hundred community solutions to global warming.

Every act of connection is an act of community. Do as many as you can. Notice that, as you actively engage the vast interdependence of life—the profound biological realities of kinship—you become better able to launch helpful actions. For the past decade, Drew and Natasha have lived next to each other in adjacent condominiums near downtown Sacramento. A while back, the two friends pitched a plan to their HOA board for approaching a nearby organic farm, to see if the farm's owners were interested in placing food-waste collection containers at their condo complex. The idea was to use food scraps from their community as compost for the farm. The owners of the farm were eager to give it a try, and the arrangement worked out beautifully. What's more, that one project got Drew and Natasha's neighbors thinking of other things they could do to support the environment. The residents who were gardeners volunteered to replace the water-hungry plants growing around the condos with xeriscaping. Today the HOA board is partnering with the local electric company, hoping to install car-charging stations in the outdoor parking areas.

Halfway across the country in Lawrence, Kansas, devoted bird-watchers Gretta and John were duly impressed when they read about how valuable outdoor time can be for helping kids deal with stress. They decided to offer kids at their church hour-long chaperoned hikes on Sunday afternoons at a nearby nature preserve. Word of the walks spread, and now three other local churches are doing the same. Two of those churches are also sponsoring kids each summer to attend a summer environmental camp in Colorado.

Focusing your actions on your community—be it your street, your neighborhood, your local school, or your workplace—will reveal pathways to change that go far beyond the latest BuzzFeed list. The fact is that any time you put your mind and body next to other minds and bodies in search of solutions, reliance on lists diminishes. Sure, there will always be a need to prioritize projects, divvy up tasks, and plan next meetings. But our real power to heal climate change is going to be fueled and guided by the creative genius of "the team." In community there is joy and purpose and hope and energy. Some social scientists call this collective intelligence. Collective intelligence is a generative version of groupthink, supporting the identification and development of innovations that wouldn't arise from the isolated thinking of one person alone.

When it comes to acts of community, don't worry about the climate deniers. Increasingly, climate change itself will provide the best persuasion for just how real this problem is. Act locally, and as energetically as you can. When the more passive or overwhelmed among us witness truthful and committed action, many will come to want the same sense of purpose and relief that comes from it. Not everyone, of course, but enough.

What's more, don't make the mistake of discounting the effect of spending time with "the choir." As environmental author Kathleen Dean Moore reminds us, the choir "makes up at least 70 percent of humanity. That's a lot!" Besides, it never hurts to stoke the spirits of those who are well engaged, thereby helping all of you to keep moving forward. We were recently reminded of this by two women in Montana, Lindsey and Becca, who got permission to plant trees in one of their city's parks. "Of course, there's value in the trees themselves," Lindsey explains, referring to their ability to sequester carbon. "But on lots of days, people walking by will stop and talk with us. Every one of those conversations is a chance to talk about climate change and what each of us can do."

Take a lesson from Lindsey and Becca: Create acts of community that carry the chance to engage people who may be on the periphery of climate change activism. Make a batch of cookies to give away, putting them on the corner of a climate information table you set up at your weekly farmer's market. Be the person who texts the neighbors to invite them to a school environmental fair. Stay on the lookout—ideally with a few friends—for chances to push climate-health initiatives with your city council or county commissioners. Keep the focus local, encouraging environmental steps that in addition to addressing climate change also address local health and the economy. Join efforts to create farm-to-school programs, linking cafeterias to local agriculture. Build walking paths, start community composting programs. Initiatives like these are observable, and easy to showcase in traditional and social media. They are acts of community building that will, without fail, lead to others joining the effort.

Diversity is another key to strong and long-lasting acts of community. In nature, diversity is by far the best predictor of long-term resilience in an ecosystem. The forest that can withstand disruption best, be it from wildfire or flood or insect invasion, is the forest with multiple species of trees and plants. And those plants, in turn, thrive best in soil that contains a big diversity of microorganism to break down nutrients, a full array of insects to pollinate the plants, and a wide range of animals and birds to consume and distribute seeds.

Not surprisingly, diversity has also been shown to enhance an array of human endeavors; it increases the likelihood of scientists making breakthrough discoveries, and the effectiveness of corporations responding to changing market conditions. Just like in a forest or an ocean, individuals of differing backgrounds—ethnicities, genders, socioeconomic conditions, et cetera—offer a wider range of perspectives and skill sets (aka superpowers). Far from being a liability, differences among humans, as in the rest of nature, yield exponentially powerful innovations.

One of the most powerful acts of community we've had the chance to learn from took place on the southeastern Columbia Plateau. There, a diverse group of tribal leaders, agricultural landowners, and biologists engaged in a long collaboration to return salmon to the Umatilla River. A key participant in that decades-long dance was Chairman Antone Minthorn, a Cayuse tribal member and elected leader of the Confederated Tribes of the Umatilla Indian Reservation. The vision the Chairman pursued was rooted in patience. It involved creating repeated opportunities for stakeholders to come together, slowing to reflect and then making incremental adjustments to ranching, logging, and farming practices on the plains of the Umatilla River basin.

Chairman Minthorn knew time was his ally. He gave years to building respectful and sincere relationships with regional ranchers, with state and national fisheries biologists, and with neighboring tribes. Those years resulted in a greater ability to listen to each other. Why did the return of the salmon matter? What would it take? What would it look like? What natural and economic impacts were important to consider? The group reviewed irrigation needs and water quality. They studied water temperature and grazing patterns. Slowly, working together, the people of the Umatilla River basin were finally able to lower the temperature of the river just enough for the fish to return.

First came just one salmon. "Man, it was rugged, beat up," Chairman Minthorn said as he and Mary walked along the river. He explained the heroic effort the fish had made to cross over the many dams between the Pacific Ocean and the Umatilla River. Then, still more salmon came, returning to their traditional home waters. "It was a miracle," the Chairman told Mary. "To all of us." Known as a sacred first food of the tribes, the salmon were also valued by the cattle ranchers, who came to see them as a sort of canary in the coalmine, reflecting the health and quantity of the water they depend on to grow the grass that feeds their livestock.

In short, this success story from the Columbia River Basin happened because a diverse group of people came together around a common value, and then used their varied skills and experiences to take action. This is intelligent leadership, drawing directly on relationship, on interdependence. It bears out predictions made by people as varied as the Dali Lama, Jane Goodall, and Stephen Colbert: that the leadership of the future will not be charismatic individuals, but groups.

BEAUTY

The hectic pace of daily life, side by side with growing concerns about our wounded planet, can make it pretty hard to take in the extraordinary beauty of the natural world. And yet that beauty can move us, can carry us, even for a little while, beyond the pervasive *me story*. Actually, this applies to most any experience of beauty. Every time you look at a magnificent painting, or hear a wonderful poem or a piece of music that deeply moves you, for example, you're touching life—your own life, the life of the artist, and the countless aspects of life held by that work of art.

Beauty is a flame in the dark, leading us back to our place within the whole of ecology. Beauty gets us right-sized. And when we're right-sized, what often comes clearer into consciousness are humility and gratitude—two qualities essential to forging creative solutions to life's most pressing problems. In the presence of beauty, people report a kind of dispositional relief. In other words, along with increased attention to things they're grateful for, they often describe comfort with their limitations, their finitude, with the fact of uncertainty. These shifts equate with feeling less embattled, less stressed. Acts of beauty call attention to what is beautiful in the natural world, all the while protecting and enhancing it. They're about making beauty a priority.

You've likely experienced the way that encounters with beauty can lift you above anxiety, taking you beyond entrenched, stubborn orientations to *me* and *mine*. See if you can experience this right now. Turn your attention, even if just for a few minutes, to nature's beauty—a bird on a branch, a cloud formation, raindrops sparkling on the tips of leaves, the deep red velvet of a rose, the way the sweeping branches of an old maple appear to hold up the sky. Notice how you're not grasping at beauty, wanting more from it, but quietly letting it take you where it will. Notice how there's no need to change it or own it.

Acts of beauty activate your heart. They prompt your natural yearning not just to belong, but to contribute to something bigger. Beauty is a connecting experience, a joining together rather than a breaking into parts. Acts of beauty are how we honor both stunning magnificence and graceful simplicity, the vast and the delicate. Here once again is the astonishment you felt for the Earth as a child, this time guiding respectful actions of restoration and repair.

What aspects of nature do you find especially beautiful? A wild-flower garden? An ocean shoreline? A redwood forest? A city park with grandmother oaks? While the health of all of nature can be said to have practical functions—from the oxygen produced by the phyto-plankton in the sea to the carbon stored by those trees—what actions might you take on their behalf simply because their beauty has made your life somehow more exquisite? What acts of care could you engage in to honor the grace they bring to your world? When you honor nature's beauty you recharge your batteries in ways that help you keep going, even in the face of overwhelming circumstances.

One day a week for the past three years, Robert, a computer pro-grammer in Portland, Oregon, has used part of his lunch hour to pick up trash from one of his favorite woodland walks in Washington Park. "There's not much litter out there, really. And to be honest, it's not so much about saving the planet. It's just my way to thank a little piece of the forest." After years of feeling nurtured by this in-city park, Robert picks up trash. It's his quiet way to nurture back. "I've always found comfort in those woods. Really, since grade school. This is about emotional connection."

Likewise, Janine, a high school art teacher in Cincinnati, started offering free *plein air* painting lessons once a month on the banks of the Ohio River. "Of course, I help people with their art," she explains. "But I started the program to bring people outside so they can learn to look. To see nature in a way that opens their hearts." Janine says that seeing the beautiful in the natural world opens you up to loving it. "It's amazing to hear painters saying they now want to do something to protect it. That was my hope all along."

In an act of beauty crossed with a bit of engineering, Glen, a retired schoolteacher in south-central Texas, found a way to divert excess water from air conditioning units in his apartment complex to supply small drinking pools for the cardinals and buntings and yellow-cheeked warblers. His lifelong fascination with birds inspired him to start inventing—serving both the birds and a lot of his neighbors, who now delight in watching the gatherings, too.

In the past decade especially, urban gardening across the country has brought nature's beauty close to home in ways both aesthetic and practical. In Detroit alone, there are now nearly 1,400 such patches of beauty scattered throughout the neighborhoods. From late spring to midautumn, these gardens yield ample food, and often flowers, while at the same time providing opportunities for social connection and environmental action. One of the most successful has been the D-Town Farm, created through the Detroit Black Community Food Security Network. It's an urban wonderland of flowers, fruits, and vegetables. The bushels of lush produce grown at D-Town are sold at local farmers' markets as well as to wholesale customers, including dozens of restaurants. Forty to fifty volunteers show up spring through fall to make it work. The executive director of the network describes one of the farm's essential yields as beauty—the beauty of good food, alongside the beauty of people interacting with the land for both environmental and economic justice.

Here in Montana, public acts of beauty are showing up too. Ranchers and other private landowners are putting their property into conservation easements: legally binding designations protecting the land from development in perpetuity. Some conservation easements preserve wildlife habitat; others protect viewsheds. Some are small, while others, like those created by media mogul Ted Turner, are very large. Our friends Bob and Annie Graham established an easement on their property along the Madison River, then joined efforts with the National Forest Service to abolish a private land boundary that was interfering with wildlife migration. Thanks to landowners like the Grahams and so many others, more than 2.5 million acres of Montana—an area larger than Yellowstone National Park—have already been preserved.

About an hour north of Yellowstone, working hand in hand with Montana ranchers and farmers, Lill Erickson founded the Western Sustainability Exchange (WSE). For twenty-five years she's been thoroughly committed to "the beautiful notion of preserving the beauty of this place." The "place" she's talking about has grown to include 1.45 million acres of regenerated prairie; in other words, prairie that's being returned to its naturally thriving weave of native vegetation. "It's growing every day," Lill says.

Yes, the mere sight of cattle grazing on sprawling rangelands with snow-capped mountains in the background is beautiful. But the beauty of regenerating the prairie goes a lot further. Even a casual glance shows what was formerly bare soil now thatched with carpets of native grass. Dig down a couple of feet and you'll find more beauty in the extra water those native grasses are able to hold in the soil; and also, in the enhanced nutrition they offer those grazing cattle. Finally, there's a great deal of beauty in the vastly superior ability of those native grasses to sequester carbon in their deep roots, thereby helping repair the unraveling climate.

Just around the horn of the Beartooth Mountains, still more acts of beauty dot the land—none more astonishing than the Tippet Rise Art Center. Founded on the principle that "Everything is involved in everything else," Tippet Rise is located on twelve thousand acres in the northeast corner of the Greater Yellowstone Ecosystem. Doubling as a working sheep and cattle ranch, this breathtaking sweep of land celebrates the inextricable link between art, nature, and the experience of being human. Last summer we had the chance to hike across much of Tippet Rise. The trails wove through rolling foothills blanketed in native prairie grasses and, in key locations, accented with massive art sculptures, each one commissioned as an expression of the infinite dance of mountains, prairie, sky, and wind. Tippet Rise, say the founders Peter and Cathy Halstead, is "where art, music, land, sky, and poetry can weave together into an algorithm greater than the sum of its parts."

Most of us don't have the resources to pull off something like Tippet Rise. Thank goodness then, for the Halsteads. They model giving back, inviting all of us into our own acts of beauty. As do the gardeners of Detroit and the cattle ranchers of the Montana prairie. Arguably, every one of these acts of beauty are also acts of community.

Check out and become an active part of any of the thousands of public acts of beauty going on right now. From the Save the Redwoods League in San Francisco to the Louisville, Kentucky Community Gardens; from the precious bird sanctuaries of urban Chicago to the hundreds of green schoolyards being created across the country through the Children and Nature Network.

Together, the search images of beauty and community may also lead you to do your own gardening, or to display art inspired by nature on your front door or the windows of your office. You might commit to sharing one beautiful aspect of nature with your children every week—a full moon, a tree coming into leaf, bees being drawn into the cups of flowers, bright birds eating at the feeder.

Any of these actions can make you and those with you catch your breath and fall in love. Acts of honoring nature's beauty open your sense of wonder and can spark you into community activism. They also serve as major nourishment for any of us already fighting the good fight in the face of climate change.

Find ways to let beauty blow your mind. They're as close as the first blush of color in autumn leaves, as the feisty dandelion squeezing its way into full flower through a crack in the sidewalk. Beauty brings us into relationship. And if you make yourself present for it, if you pay attention, you'll feel it igniting an urge to *reciprocate*. To give something back to this dazzling world.

MYSTERY

No act arising from interconnection occurs without some measure of mystery. Does that leave you feeling a little unsettled? Do you feel like the results of your action have to be thoroughly envisioned—fully mapped out before you even begin? We understand this. It's how we, too, were raised to think, a way we've found ourselves thinking as we've worked on writing this book. This engineered way of approaching action got ever more entrenched as we worked our way through school and into careers. It's no surprise that the challenges of climate change prompt us to overfocus on the hard numbers: parts per million of carbon in the atmosphere, degrees of warming per decade, rates of glacial melt. These facts *are* important. Yet when it comes to you taking personal action to mitigate and even heal climate change, holding a place for mystery—making peace with the unknown, and *cooperating* with it—will keep you both sane and resilient.

Mystery takes off our masks. We're left with exactly who we are, here in the uncertainty we can never avoid. You already understand pretty well that no matter what plans you might make for any given day, you just can't know how things will go. You also know that if you're counting on things happening in a particular order with particular outcomes, well, by day's end you're likely to be pretty cranky. Yes, there are things we can count on: spring follows winter, the atmosphere continues to provide the air you need to breathe, and for the rest of your life gravity will keep your feet on the ground. But much of what happens on this planet is improvisation. On a random summer morning, pine bark beetles suddenly swarm into a healthy lodgepole forest looking to lay their eggs. The trees respond in the moment, squirting sap out of every hole the beetles drill into their trunks. If the trees are lucky, they'll be able to foil those beetles, pushing them off to somewhere else.

Act and adjust. Uncertainty is the name of the game, and improvisation is the way to roll. The way you get good at improvisation is to be deeply in touch with what's actually going on around you (instead of rehashing what's past or rehearsing for what *might* happen later). Remember, nature has wired you for the right response right now. This latter point especially—that you already have what you need—absolutely shimmers with mystery. Indeed, this is one important facet of climate change: it demands in no uncertain terms that we trust our bone-deep intuitions.

You've likely relied on that kind of trust more often than you know; yet for most of us such trust is underdeveloped. Accepting the mysterious, welcoming the flow of the uncertain, is the most effective way we know to sharpen your intuition. Making friends with mystery leads to better discernment, because it frees you from making decisions based entirely on the limits of your intellect. If you flinch at that, keep in mind that no less an intellect than Albert Einstein advised his students that if they had to make a choice between knowledge and mystery, they should always choose mystery. We think and plan, we ready ourselves for improvisation, and we trust.

The mystery we're talking about is the way life maximizes the opportunities held in any given moment. With mystery, we can synchronize our lives with the lives around us to grow in a way that fits seamlessly with current circumstance. At just the right time, buds swell and come into bloom. Almost overnight, spring rains draw overwhelming flushes of green out of winter-weary prairies. Young birds fly migration paths, some without their parents, knowing the way even though they've never made the trip before. The natural world is sustained with a blend of spontaneous collaborations, unforeseen sieges, and resultant acts of repair.

What makes all this rising and falling, ebbing and flowing possible in the first place is another, perhaps even less comfortable, kind of mystery—something we've come to think of as nature's *loving disinterest*. Loving disinterest is causeless and ineffable. It's love in the way of the ancient Greek notion of *agape*, big and mysterious and constant and unconditional. Each and every day, we awaken to an abundance of the essential ingredients that sustain our lives—no questions asked, no tests to pass, no performance required, and no investment one way or another in whether we notice. To our minds, the only term large enough to hold this profusion of mystery—simultaneously neutral and boundlessly generous—is *loving disinterest*.

Gary has walked more than thirty thousand miles through wild nature—now and then alone, and often in the company of wildlife. Through mountain meadows and along unbound rivers, in deserts and rainforests and across the grassy savannas of East Africa. Across those miles, nature's loving disinterest, whether noticed by Gary or not, has been a constant. No matter how verdant or barren the landscape, the natural world keeps providing for those who live there. Life is continually reaching for more life, more variety. Unfolding with deeply natural, yet altogether unfathomable, logic.

By careful design you are a deeply centered being, with highly evolved instincts and reflective skills. Your best act in any given moment arises from within a peaceful relationship to uncertainty—which will always call forth your broadest and most relational knowing. Conditions change all the time, sometimes catching you unaware. And sometimes you screw up. But that doesn't mean the guiding instincts that are a part of your human nature have gone anywhere at all.

To nature it makes absolutely no difference the kind of car you drive, the cost of the clothes on your back, or where you went to school. In that way, and in very practical terms, the natural world functions as a kind of leveling force. Way back in the 1700s, that particular aspect of nature led some newcomers to North America to describe wild nature as "the great equalizer." Later, small icons of that nature—leaves, acorns, birds, pinecones, even a national tree—became powerful symbols in the effort to launch American democracy. Unfettered nature, in other words, and in particular its unmerited gifts, became emblematic of patriotism.

Growing our awareness of loving disinterest encourages our ecological intelligence. You'll find the truth of that reflected in the words and actions of true elders, who are arguably the most ecologically intelligent among us. In true elderhood, after all, relational thinking gives birth to acts of benevolence that take into account the well-being of *all* involved, now and into the future—without any need for thanks, or even notice. To be fully wise, in other words, is to be aligned with the fundamental nature that resides in us all.

The rapid, shocking spread of COVID-19 early in 2020 had no shortage of mystery. Certainly, it cued panic. For some, that panic showed up in hoarding hand sanitizer and toilet paper. For others it led to massive stock liquidation and layoffs. All along, however, there was evidence of human nature at its best, of ecological intelligence signified by great courage and empathy. These natural responses stood throughout as enormously positive reactions to the mystery of this pandemic.

The latest evolutionary biology has much to say about how the evolution of species contains "isomorphisms." In other words, what happens on the large scale tends to be replicated on smaller scales too. Take, for example, your immune system. First of all, it's ancient, having been under construction and refinement for millions of years. It's the system in your body that, when alerted to the intrusion of a hostile outsider, like coronavirus, activates processes for getting rid of it. Your immune system tags the invader and calls in powerful infection-fighting cells—even generates the kind of fever that helps the good cells do their job better, while subverting the virus. Most of the time, these responses are successful. You feel crummy for a couple of weeks, and then you recover.

This parallels processes of evolution over the long haul. We now know that Herbert Spencer's twist on Darwin's words, describing evolution as survival of the strongest, toughest *individuals*, was off the mark. Rather, like the collaborative communities of cells that make up your immune system, evolution is far better described as survival of the friendliest—the most cooperative. In fact, evolutionary theory attributes human success to our having consistently created highly cooperative *groups*. The constant across all of the challenges the human species has faced is the presence of other people—our social ecologies. For reasons we can't fully understand, your body's immune system operates at peak levels when you're socially connected.

The way that cooperation unfolds, how it's unfolding even now, is a bit mysterious. Think about how with coronavirus, restoring health relies on people acting from interdependence. We simply have no other choice—that is, if we want to live, or at least avoid a great deal of suffering. The panic and disharmony set loose with COVID-19 arose from the same old separation focus—*what about me and mine?* But in short order it became clear to many of us that we are only as safe as everyone else. We're entirely connected. Even the calls to isolate grow out of interconnection.

We survive with and because of other humans, all of us rooted and nurtured in one constant: the natural world. In cases of public health, which can be understood as our collective immunity, it can take a minute to orient and then act for what the situation calls for. Just like the immune systems of our bodies, our human communities are outfitted to respond to intrusive circumstances—be it COVID-19 or the crises of climate breakdown. As any biologist, immunologist, or even historian will tell you, a successful response requires staying open to mystery, to uncertainty. There are no lone heroes. There are no silver bullets. Instead, we must continually stop and ask to discern our best next actions. And always there is mystery.

Here's the thing: Nature has its way. To humans, it's mysterious and uncertain. But of course, uncertainty doesn't bother nature. Nor does it bother the truest scientists. This is because these researchers are truly awed by any opportunity to use their thinking to learn more and more about how the natural world works. They engage in acts of mystery with every investigation, knowing full well they will never know it all.

The COVID-19 pandemic is in many ways a harbinger of the changes coming with climate breakdown, insofar as they too can't be fully understood in advance. Still, in the coming years we can either take every step possible to restore the health of our planetary system, or we can persist in acting out of the illusion that we're separate from the natural world that birthed and sustains us. The path before us is ours to discern. Nature will advance what is sustainable and will let go what is beyond repair.

You have the opportunity to make friends with mystery. If you do, the actions you take will better align with the well-being of our interconnected world. Whether you're recycling or practicing regenerative ranching, you can make a difference. Everything you do can contribute to the patient reordering and renewal of real-life health and well-being.

None of us knows how things will turn out. Floodplains devastated in one season might in the next sprout vast runs of grasses and flowers. Meteor strikes happen, and so do hundredth monkeys. Acts of mystery mean showing up in the unpredictability of it all and doing our best. A best that is deeply rooted in our human nature.

INSPIRE

Inspiration grows when you start watching for it. It takes attention and practice to act consciously with and within the natural world. But all along the way, the presence of inspiration—being inspired and being inspiring—is a sure sign of sustainable social ecology, an alignment that supports the well-being of all life. To inspire and to *be* inspired come from honoring relationships. In this fourth practice of Full Ecology, we are, at the same time, culminating and regenerating the whole process.

Inspiration follows from stopping, asking, and taking actions of repair. Expressed naturally, inspiration happens without need for excess energy or applause; that is, without performing. Inspiration shows up when you take full responsibility for your relationship with the whole of nature, including your own close-in social networks—your communities, your neighbors, your intimates; and most local of all, your physical, emotional, and spiritual self.

The benefits of paying attention to relationship show up throughout the natural world. Even our less immediate relationships with nature reveal guidance. In her book *Timefulness*, geologist Marcia

Bjornerud invites readers to imagine the speed of rocks—the speed of mountains and ocean bottoms. She calls on us to take a closer look at the observable evidence of earth, fire, water, wind, and gravity stretching across billions of years. Such mysteries bring us right to the heart of loving disinterest. Nature unfolds as it unfolds, when it unfolds.

The idea of timefulness draws us into remembering that our beautiful home planet was formed by staggeringly patient and coordinated processes over billions of years. Think of how long it takes for rain and wind to wear away a whole mountain range. Or cast your imagination far, far back into prehistory, when the only living things in the whole world were in the sea. Such consideration places our acts of consciousness and kinship squarely in the arms of mystery.

Given the impact humans are having on the natural systems of our planet, it's no surprise that some scholars have taken to calling our current geologic epoch the Anthropocene. As writer Robert McFarlane suggests, "As the Pleistocene was defined by the action of ice, so the Anthropocene is seen to be defined by the action of anthropos: human beings, shaping the Earth at a global scale." We can see how this term might be helpful for getting people's attention, for aiming us toward healing rather than harming. But we also know it holds the risk of sliding once again into making everything all about us humans.

Yes, we're having tremendous impact. And yes, it's not just urgent, but a matter of life and death for all kinds of beings that we reduce our harmful impact on the planet that supports us. But we humans are only and always one small thread of the mysterious web of creation. Taking inspiration from the larger mystery that contains us all, and acting from there, is a great way to move ourselves out of the center of the story. It's just not all about us. And that can be a big relief.

The Earth will make it one way or another—likely until the sun begins to die. That's a long time. As long as we humans are here, we have the option of being companions, stewards, grateful participants in what is still, and always will be, an Earth-centered epoch.

So, with a little practice, you'll find you have a few startling glimpses of time itself as rocks and rivers and mountains. From the platform of separation, this view engenders that old and awful feeling of being frightfully small, or insignificant. But from connection, it reveals the magnificence within which your life unfolds. Ours is an endlessly creative planet and has been for more than four billion years. It didn't just give you a fact-driven brain, but rather wired you with extraordinary capacities for wonder—for kinship and generosity and love.

Exploring and growing those capacities is at the heart of inspiration. Generosity and love move us to take into account even the lives of those who will come along after we're gone. This "being a good ancestor," as Jonas Salk once put it, can only come from living out of care each day without any need of notice—being a source of inspiration, nonetheless. Inspiration that follows naturally from living with loving attention rooted in the soil of a mysterious future. We reduce our toxic waste. We tend the health of the atmosphere. We pay forward to future generations a level of vitality that preserves for them the chance to live and act in highly creative and inspired ways.

This part is simpler than you might think. Every time you stop, ask, and act from your wisest self, others will be inspired to do the same. This is the lifeblood of any chance we have to heal the planet. And you are totally up to the task. We know such claims can be hard to believe. You're a person with scars and bruises and foibles and regrets. Like most of us, you could fill a good-sized cargo ship with your mistakes. Yet for all your flaws and clumsiness—and in many ways because of them—you inspire by continuing to show up. You've never let go of believing mightily in this planet, and in all the beings who call it home. Each one worth saving.

We agree.

And living out of that truth is the most important thing any of us can do.

Over fifty years ago, long before we were able to grasp the realities of climate change, the monastic scholar Thomas Merton described the way western societies had abandoned inspiration, and thus themselves, adopting instead the ways of crusade and conquest. These ways were crudely simple. Their blustery promises of certainty, of control, proved hard to resist. We traded away the ground of intimacy, turned our backs on our fellow travelers, and set off alone to claim our fortune. But the richest life, the inspired life, as Merton suggests, is actually all about intimacy. "A deep resonance of one's entire being...set up with the entire being of the other." Heart speaking to heart.

We all know that we're well into great climatic upheaval. The flooding of farms in the Midwest and cities like Houston and New Orleans; the terrifying megafires that consume too much of California, Colorado, and southeastern Australia; the ferocious hurricanes; and the severe droughts will devastate still more towns and cities, still more farms and ranches in the years to come. There's no playbook for what we're facing. And while there are cleanup crews, there's no cavalry.

Crises like these are life changing. Yet it's up to us to decide what those changes will look like—what part of our humanity they will feed, whether our innate compassion will be diminished or strengthened.

The ability to share hope and heart is a highly adaptive human quality. It's something that first responders see all the time. Tragic circumstances spark inspired acts of community, and with notable efficiency. Following a devastating wildfire, the pharmacist and the plumber stand side by side picking through rubble, listening and look-ing for survivors. A tornado touches down, and soon the student nurse and the accountant are tending to the elderly, the injured. In the wake of hurricane flooding, the carpenter and the lawyer find themselves in a rowboat together, rescuing people from rooftops, while members of the high school football team hand out bottled water they found stored in the gym.

These responses to crisis involve physical movement and energy. Lots of it. Bessel van der Kolk, a psychiatrist who pioneered body-centered therapies for trauma recovery, describes the evolutionary advantage held in this ability of humans to show up. In the midst of catastrophe, engaging the body in energetic activity has immediate attenuating effects on the nervous system, preventing much of the deep emotional scarring that can come with trauma. At the same time, for those disabled by a crisis, the knowledge that helpers are supporting them sets the needle toward recovery. Quite beyond the usual fixation with *me stories*, the scene in the immediate aftermath of disaster is one of supremely interconnected human nature.

Of course, when it comes to climate change, we can imagine the worst. But we don't have time to get stuck. Worry might always ride along with us, but it's time to take it out of the driver's seat. We can acknowledge and hold our anxiety more lightly even as we pull together to act with calm and conscious attention. This is what *inspire* looks like.

Watch the people who inspire you most. Watch for all the times they act without announcement, without self-focused progress reports. One of the reasons you're drawn to them is that rather than just talking about useful behavior, they're out there living it. Over the years, the two of us have come to see that humanity's treasured guidelines for how to live, from the Ten Commandments to the Eightfold Path, didn't just show up in books or on tablets of stone. Instead they were observational notes taken by onlookers to document living truth: what it looks like when people live fully aligned with their human nature.

One of the best places to see this is in the lives of true elders. These are typically people with many years and much experience, whose unselfconscious clarity and easy authenticity is as comforting as it is inspiring. A true elder's first thought is for the well-being of the most. Motives for fame and fortune, if they were ever present, have left the stage. Material reward and individual recognition have become less interesting. The humility of elders isn't something performed. It's just what full kinship looks like. It's human nature at its most integrated. And at its most inspiring.

When Gary was writing about those amazing teens in wilderness therapy, the education director for that program was a sixty-three-year-old named LaVoy Tolbert. To Gary, LaVoy was the essence of elderhood—not just because of what he knew, which was plenty, but because of how well he listened. By simply getting out of the way, he really heard, saw, and even felt what was going on. He didn't try to fix things. Instead he stood a bit off to the side, and from there helped you see how you already held your own best answers.

Which is how it happened with the family of a sixteen-year-old girl from California named Jamie. At the end of eight weeks living in the backcountry, students reunited with their parents in an emotional ceremony at the edge of the wilderness. Most kids were nervous but excited to see their families. Some, though, like Jamie, were frightened. After slowly walking into the clearing where everyone had gathered, she stopped six feet shy of her parents and broke down in tears. Then she turned around and walked off, dropping to her knees at the edge of the circle in the shade of an old ponderosa pine.

Her father looked heartbroken. "Maybe we shouldn't have come," he said to LaVoy. "Do you think we should just go home?"

LaVoy put his hand on the man's shoulder and looked at him tenderly, eye to eye.

"She just needs a little time," he said. "You've probably been in a situation where you decided something was right, and the next day you knew that wasn't the move you should have made."

"Yeah," her father whispered, near tears.

"Well, then," LaVoy said. "You just relax. Stay right here. There's nothing in this world that can stand up to the force of love. You give her all the love you possibly can. Tomorrow's another day, and things will turn around."

Out in the wilds, Jamie had lived through the powerful experience of coming into her own. Part of her fear of reuniting with her family was that she might lose touch with that, opening the door for the old troubled and addicted Jamie to take over again. Yet what LaVoy told her father, about the force of love, he'd also told Jamie the day before. And sure enough, early the next morning, father and daughter did find themselves reaching out—across the history they shared, across the vast distance between their ages, between parent and child; each with enough love to begin trusting the love of the other. There, they found that together they had more strength, more insight, more courage and belonging than they'd ever had when they were standing alone.

Elderhood in socially oriented mammals, from humans to bonobos to elephants to dolphins, rests in an exquisite balance of relationship and agency. As for the agency part—*doing* something—actions of elders are well considered, and then highly efficient. At the same time, they always reflect a sphere of interrelatedness. This isn't to say that a human elder's character flaws, oddities of temperament, lingering worries, and obsessions don't come along for the ride. We're not talking Glenda the Good Witch, here. More like Dumbledore with a twist of

Mad-Eye Moody. Elders provide inspiration as fully natural beings, warts and all. And here's the really good news: just like in the rest of nature—because we *are* the natural world, after all—the capacity for elderhood in humans can't be squelched.

It's probably true that elderhood is most available to those who have lived long lives, seasoned by success and failure, joy and loss; people who have had plenty of chances to explore the values of generosity and cooperation. Yet no matter your age, you carry inside the *sensibility* of elders. In other words, you already have an "inner elder" all your own. This is what allows you to recognize elder wisdom when you see it, and why you feel drawn to it. Your inner elder is your natural inclination for relational living—for action rooted in strong connection. This inclination is deep in your bones. And as with other genetic traits, it's your encounters with life itself that keep offering chances for these traits to be expressed.

This natural genius is yours for good. It can never be eliminated as a possibility. The natural world might sometimes appear to be conquered: dams bring mighty rivers to a halt, forests are clear-cut, fields of diverse plant life are plowed under and planted with a single fragile crop. Yet over time the dam weakens as the rock and soil subsides, as the water continues to pluck at the retaining wall. And no matter how many centuries it may take, the forest and the plowed field remain ever at the ready to resume their uprising. The same is true for you. You can be indoctrinated in separation theologies as a youngster. You can be subjected to an unending stream of ridicule for your sensitivity and your ideals. Yet like the river or the forest or the field, your wisest and most connected self—your inner elder—is always at the ready. As terrifying as climate change is, don't underestimate the power it holds to put that best self into action. Then watch. As you persist in living that way, deep in your truest nature, people around you will join in. After all, it's their nature too.

In this culture, it's a habit to think of inspiration in terms of the mind—in particular, a quiet, unruffled mind. Decades ago, when we didn't yet know one another, each of us spent a fair bit of time exploring inspiration from that very assumption. Each on our own, we picked up what even today remains a common misunderstanding: that the body with all its needs is a barrier to awakened consciousness. The idea was to separate our bodily experience—what Saint-Exupery's *Little Prince* called "matters of consequence"—from the part of us that yearns for something more. Clear awareness, the thinking went, can only be gained by turning our attention away from everyday concerns: bills to pay, kids to raise, cars to repair, groceries to buy, successes and failures to monitor and compare. Inspiration was to be found by focusing on the infinite rather than the finite.

Gary would sometimes spend days in reading and contemplation. On a great many occasions he'd walk alone into the wilderness, sometimes for weeks at a time. Mary, meanwhile, would take extended retreats in meditation halls or make deep dives into nature preserves—all the while, keeping silence. But the retreats would end, and reliably, the usual stuff of everyday life would pop up again. It was a shaky setup from the start. Tucked in retreat, each of us experienced sharpened awareness. But functioning in our families, at work, and in our communities required complicated engagement.

By the time we met, both of us were wrestling with this conundrum: how to bridge our everyday lives with what we'd come to know on retreat from those lives. We wondered with each other how we might live, down to the most tedious of tasks, in ways consistent with the wisdom and beauty of the present moment. It was time to check and see, when we were off the trail, off the zafu, and back into the mess of our everyday human lives: *Does that real essence of being ever go anywhere?*

During a quiet thirty-minute meditation it's possible to put down a lot of anxiety and desire, which is great. But the formal practice of meditation, unlike much of the rest of life, is carefully arranged. Precisely bounded. To have any semblance of meditation-like inspiration in the many other minutes of our lives, we realized we'd have to make friends with uncertainty; with the reality of impermanence, of constant change. The impermanence of summer. The impermanence of squash or apples from this year's harvest. The impermanence of toddlers who are suddenly adults. The impermanence of grandparents and parents who pass away.

Impermanence calls us back to *life in a body.* There's no getting around the fact that human nature is embodied. The holy, the true, the real—by whatever name—can only be known here. Locally. In our bodies, and in whatever everyday matters of consequence we face.

The great thing about inspiration is that it lives in pretty much everything, including a lot you never pay attention to, and a lot that you hold at arm's length. Searching for a title for one of his books, Franciscan friar Richard Rohr settled on this one: *Everything Belongs.* If something is here, it belongs. Nothing can be ruled out. Rohr was hardly the first to make this conclusion. Early in the thirteenth century, Rohr's spiritual forefather, Saint Francis of Assisi, centered his ministry on being in full relationship with the world. In that sense we, too, as part of Full Ecology, are just two more people in a long line who see belonging as a cornerstone of vital living. Everything you see matters, including the people, things, and conditions you reject—even those you cannot stand. If you really want to reinhabit your earthly experience, consider that everything on the planet belongs.

The lessons of nature can't be picked like cherries. We don't get to stuff our pockets with only what's convenient, entertaining, or obviously beautiful. We have to step all the way in. Into the discomfort that comes with challenging our own conditioned reliance on being separate. And also, into the willingness to be wrong, admitting that our understandings are always incomplete. Arm in arm with that inevitable uncertainty, at peace with it, we find ourselves at last able to see the brilliance in all we encounter—people, cities, forests, mountains, rivers, weather.

Inspiration can feel irrelevant, even indulgent in the face of the real and often frightening challenges we face in climate change. In the coming days and months, make it a habit to pay close attention to what you're scared of. Maybe you're afraid for children and families who are losing their homes to floods and fire. Maybe you're scared water and food will become scarce. Maybe you're afraid all of this is coming your way—to your community, to your loved ones. It's no wonder you're exhausted. Fear takes energy.

But if you stand quietly with that fear for a time, holding space for it, not judging or reacting to it, pretty soon you'll see that there's inspiration in the borderlands linking your circumstances with your fearful reaction. Stay put for a while. Keep in mind that this fear has arisen because you care. This is sacred ground. It is a place for you to watch and listen. In a very real sense, this interior landscape can be seen as what biologists call an ecotone—the place where two ecological systems come together, and a place particularly rich in possibility. Like along a shoreline. Or where a meadow meets a forest. Or the marshy wetland that lies between the flow of a river and its outer banks. The life systems that come together in the margins make possible an intermingling of species. This yields extraordinary diversity, vibrant abundance, and uncommon innovation.

Each of us lives every day along these kinds of edges. Ecotones of family, where one generation meets another. Ecotones of personal identity, where aspects of who you are intersect with aspects of others. Ecotones of gender and vocation, ethnicity and sexual orientation. Ecotones of community, where different cultures, economic classes, and political leanings are as common as weather. And on a more basic level still, ecotones where your stories and your circumstances collide.

Fear itself is one such ecotone. Throughout the day, variations in circumstance cause natural physical reactions to arise in your body. Sometimes your thinking kicks in to construct a story. The story is especially likely to catch your attention when it's one of danger, or disrespect, or attack. The circumstances might not be as problematic as they seem, but the stories render them so. Out in the woods, you jump to the side of a trail certain you've spied a snake when, on inspection, you see it's only a branch. You're offended by a colleague who seems to have given you the side-eye; but had you listened a little longer, you'd have heard about the dust that flew into her face just before the two of you started talking.

It may not seem very palatable, but this unsettled ground actually offers profound opportunity to repair your own internal ecology. With patience and attention, you can come to discover the power of holding still.

Here you make a full return to *stop*. Right in the middle of your circumstances, you put down every story—fearful or not. Free of story, you will see what is true. This, then, is the landscape of inspiration.

Inspiration is an overlooked treasure of being human. It's an essential source of energy that can persist through disruption and confusion. Human nature generates inspiration not just through expressions of beauty and joy, but also by holding despair and even dread with open palms. By being vulnerable. By walking ever forward through weather fair and foul, open to it all—every animal and plant, every earthquake and storm.

In his book *Being Mortal*, practicing surgeon Atul Gawande reflects on inspiration by considering what might seem an unsettling question: What makes for a good death? His conclusion: Live all the way to the end. Full Ecology might describe this as embodying your nature at full capacity. Like a tree, or a frog, or a tiger. Living in all of who you are, all the way. So, for example, think of today. What does a good day look like? We suggest that it looks very much like a good life: twenty-four hours of living fully. That includes living well with loss and grief.

The stronger your relationship with inspiration gets—both as giver and receiver—the less threatening loss will become. Inspiration is how you remember the world has your back. Without planning or intention, the universe unfolds, giving you what you need even in the hardest times, in ways that are supremely reliable. You can count on this. The fact that we can't always understand it changes nothing. Accepting your life as an expression of natural logic doesn't remove the hardships, but it leaves you always within reach of the joy of being a full participant in the whole of life, no matter what circumstances hover on the horizon.

Much like the climate scientists embracing their fear and despair while gathering data that show what's heading our way, your willingness to embrace fear and despair while you act allows your concern for the world to be more powerful, more public. Opening yourself to grief instead of running from it will always reveal inspiration. It will restore your sense of belonging. It will deepen the color and texture of your relationships with family and friends, with the Earth, with life itself.

What you're signing on for is the kind of growth that is the natural way of all things. Upheaval is followed by strengthening and recovery. With fresh vitality comes a new order of things, only to be jostled again by disruption. Some of these cycles will be tiny: Slicing your thumb while cutting vegetables. Cleaning the cut and applying Neosporin and a Band-Aid. Returning to the cutting board a bit more attentive. Other cycles will be big, and frightening: Losing a loved one. Losing your home or your marriage, your job or your health.

The growing demand to meet the challenges of climate change fits into the category of *big*. Every action that any of us takes or doesn't take sews a seed that will root and grow across centuries.

Climate change places us in a particular version of mourning that some psychologists refer to as ambiguous grief. This is what you feel when what you're grieving is still present in your life. The natural world, for example. The rivers, the bears, the air, the children, the infants newly born this month. Our grief is not something to run from or to think of as pathological. It's life itself.

Knowing that, we walk with these wounds as our traveling companions. They need not hobble us nor make us less effective. In fact, grief itself offers enduring inspiration and a liveliness too little experienced—the liveliness of a planet twirling with truth, with beauty, with community and mystery. So, we travel on, resilient even as we reconcile our hearts to what poet Stanley Kunitz calls "this feast of losses."

Repairing our climate cannot be accomplished in our lifetimes.

Indeed, the work will almost surely extend beyond the lifetimes of every person currently on Earth. Like the Duomo of Florence that required the best efforts of six generations of builders, ours is a project to which each of us contributes in the time we have. Quite unlike a duomo, however, our efforts are a matter of life and death for millions of beings. We, uniquely, face an urgent call to give our best in the interest of our own lives and the lives of all that follow.

Whether you know inspiration as a constant flow or as seemingly random flashes, it's here for the duration. The reliable center. It can appear to come and go like so many butterflies, but such is the nature of human attention. Ancestral stories from groups as varied as the Navajo and the Irish remind us that while the center can sometimes go missing from view, it's always here.

The script is yours to author. In fact, you're already writing it—with every thought, every action—each a contribution to the larger story you're coauthoring with all your relatives, not just those who are human, but also with forests and oceans and deserts and grasslands and all the creatures they hold in their embrace. A story written together with the storms of human conflict and atmospheric disturbance, and with all the vast and beautiful spaces in between. Such is the ecology of your days. A fine golden thread shining in this wondrous weave of life on Earth.

ACKNOWLEDGMENTS

We are deeply grateful for the relationships that have led us to Full Ecology. Special and heartfelt thanks to Mayme Porter, LaVoy Tolbert, Mary Alice Kemp, Sara Henning-Stout, Chuck Ebersole, Roy Sampsel, Terry Gutkin, and Jane Conoley—each of whom has shown us the very best of human nature.

In addition, our thanks to Steve Wasserman at Heyday for his vision, and to editor Marthine Satris and her team for their brilliance in helping shape this book.

Finally, our profound gratitude for this planet and the wise guidance it holds for us all.

ABOUT THE AUTHORS

Mary M. Clare and Gary Ferguson have each dedicated more than thirty years to exploring the world's social and natural ecologies— Clare as a graduate professor of psychology and education, Ferguson as a nature and conservation-science writer. Their individual work has now come together in Full Ecology, a movement devoted to the idea that we can best serve both the natural and human world by reclaiming our human nature. Clare is a fellow in the American Psychological Association, and with her PhD in psychological and cultural studies she has published more than one hundred scholarly articles and two books. Ferguson is an award-winning and internationally recognized author of twenty-six books, most recently *The Carry Home* and *The Eight Master Lessons of Nature*. When not on the road or in the wilderness, the couple live in Bozeman, Montana. Find out more at fullecology.com.